JAPANESE OCCUPATION "BALL FURONG" COLLECTION
1941-1945

南方占領地のフィラテリー
玉芙蓉コレクション

Publisher: Stampedia, inc.
Editor : Tamaki MORIKAWA
Date of issue: July 1st 2018
Number of Issue : 300
Price : 2,000 Yen (VAT included)

© Copyright by Stampedia, inc.
7-8-5-902 Roppongi, Minato, Tokyo, 102-0083
Design : Takashi YOSHIDA, Tokyo
Printing : P&A, inc., Tokyo

本書の概要 / GUIDANCE

　1941年（昭和16年）12月8日の真珠湾攻撃によって始まった太平洋戦争で、日本は米英蘭をはじめとする連合国と全面戦争に突入、英領や蘭領などの資源を求めて東南アジア全域に侵攻、昭和17年3月までの間に制圧し、占領した。

　これら占領地においては、日本軍による軍政が敷かれた。郵便事業も接収され、日本軍監督の下で郵政が施行されることになった。　その時に発行された切手がいわゆる「南方占領地切手」である。南方占領地切手には、現地切手に加刷して発売された加刷切手と新たに印刷をした正刷切手とがある。

　蘭印地区を除いて概ね解明されているこの分野ではあるが、蘭印地区における加刷切手については戦後70年以上経過した現在でも未だ解明できていない部分が多く残っている。

　本書に掲載されたコレクションは、私の友人の、ある日本のフィラテリストが生涯かけて収集した「南方占領地切手」のコレクションで、戦後間もない時期から60年以上にわたり南方占領地全般（切手およびステーショナリー）を収集・整理した一大コレクションである。

　氏の収集は南方切手にとどまらず、日本切手や沖縄切手などでも多くの珍品を所有していたが、とりわけ情熱を注ぎ続けてきたのは、南方占領地のコレクションである。国内展はもとより国際展へも幾度となく出品され、数多くの賞を受賞するとともに南方占領地切手の研究に大きく貢献してきた。しかしながら、その全容を一堂に展示する機会はこれまでになく、今回ご遺族のご協力のもと、日本のナショナル・ミュージアムである郵政博物館で初めて一同に展示することができると共に、その展示内容を書籍化することができた。

　氏が収集途中であった分野においてもリーフをそのままの形で掲載し、お亡くなりになる寸前までその収集に情熱を燃やしていたことをご紹介したいと思う。また、このような記録として残すことによって今後のこの分野の研究に大きく貢献できることを期待している。

平成30年6月8日

守　川　環

目次 / INDEX

フィリピン	PHILIPPINES	1
香港	HONG KONG	13
ビルマ	BURMA	17
マライ	MALAYA	29
北ボルネオ	NORTH BORNEO	69
スマトラ	SUMATRA	77
ジャワ	JAVA	113
海軍担当地区	NAVAL CONTROL AREA	121

各地域とも概ね発行順に展示しています。

また、本書では見開き 2 ページを左から右のページへ、続いて下段へと 4 リーフごとに展示同様に配置しています。

本書を購入した「スタンペディア日本版」会員は、以下の手続きを完了することで「マイスタンペディア」にて実寸でスキャンされた全リーフの PDF を会員在籍期間中、ダウンロードできるようになります。

[手続き] 以下の電子メールをお送りください。

① 宛先：tpm@stampedia.net

② 件名：玉芙蓉 PDF ダウンロードの手続き依頼

③ 内容：住所・氏名・いつどこで本書を購入したか

＊本サービスは「スタンペディア日本版」会員限定サービスです。（年会費 2,000 円）

PHILIPPINES

1942年（昭和17年）1月2日　　　マニラ占領
1942年（昭和17年）1月3日　　　軍政部を発足
1942年（昭和17年）3月4日　　　郵便再開日。加刷切手の発行開始
1942年（昭和17年）11月12日　　現地印刷の正刷切手の発行開始日
1943年（昭和18年）4月1日　　　日本印刷の正刷切手の発行開始日
1943年（昭和18年）10月14日　　比島共和国成立（軍政廃止）
1945年（昭和20年）2月3日　　　占領郵便の停止日

PHILIPPINES

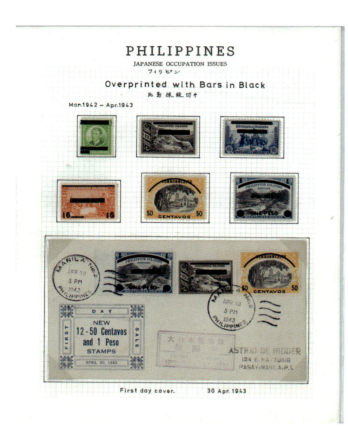

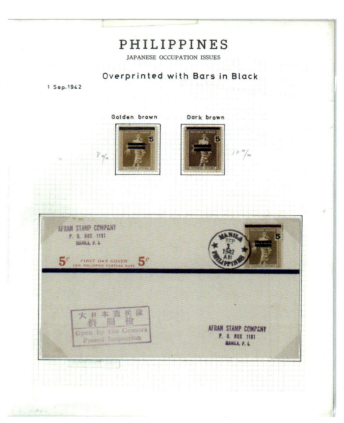

JAPANESE OCCUPATION "BALL FURONG" COLLECTION

PHILIPPINES

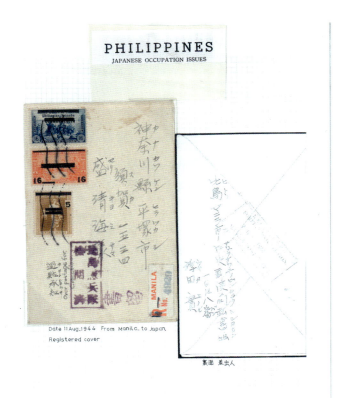

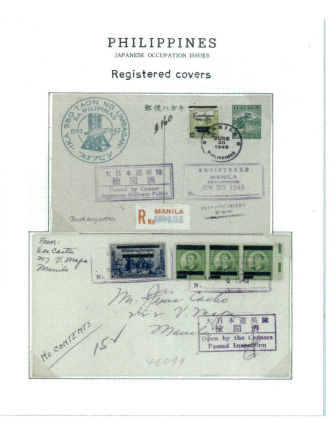

PHILIPPINES

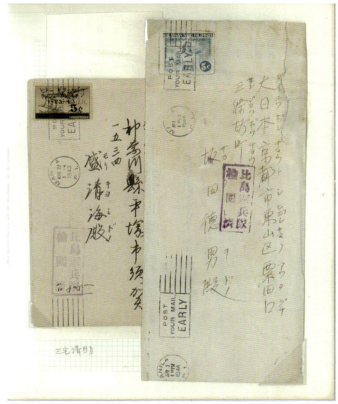

PHILIPPINES

南方占領地のフィラテリー　玉芙蓉コレクション

PHILIPPINES

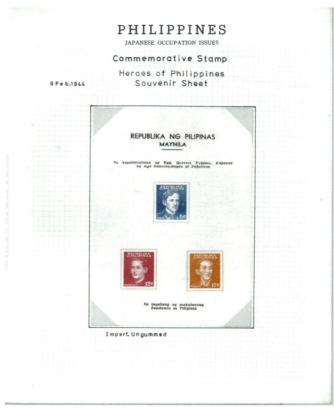

JAPANESE OCCUPATION "BALL FURONG" COLLECTION

PHILIPPINES

PHILIPPINES
JAPANESE OCCUPATION ISSUES

Commemorative Stamp
Heroes of Philippines

12¢ stamp omitted

PHILIPPINES
JAPANESE OCCUPATION ISSUES

Commemorative Stamps
Second anniv. capture of Bataan & Corregidor
ベタアン、コレヒドール 陥落 二周年記念

7 May 1944

First anniv. of Independence
独立一周年記念

12 Jan 1945

PHILIPPINES
JAPANESE OCCUPATION ISSUES

Official Stamps
Overprinted or Surcharged in Black with Bars and

公用
(K. P.)

1 Jun 1943-1944

Wider spacing between bars

F D C.   28 Aug.1944

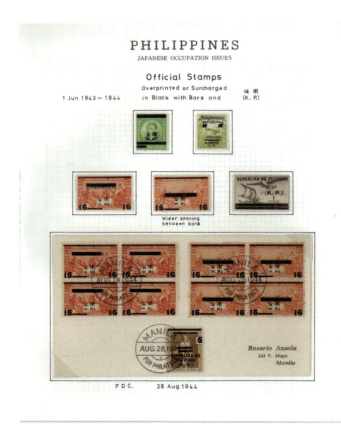

PHILIPPINES
JAPANESE OCCUPATION ISSUES

Official Stamps
5C. On 6C.

公用切手

1 Jun.1943-1944

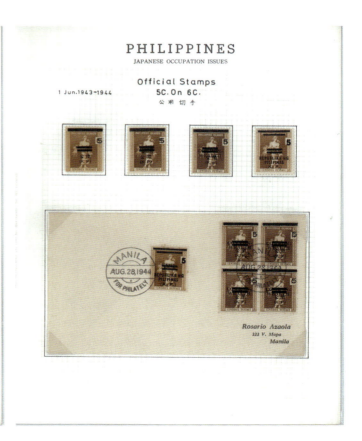

7

PHILIPPINES

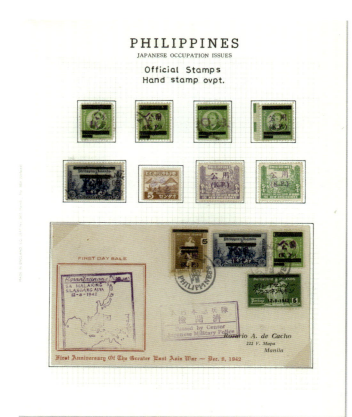

PHILIPPINES

PHILIPPINES
JAPANESE OCCUPATION ISSUES
Official Stamps
Hand stamp ovpt.

PHILIPPINES
JAPANESE OCCUPATION ISSUES
Oct. 1942
Official Stamps
Hand stamp ovpt.

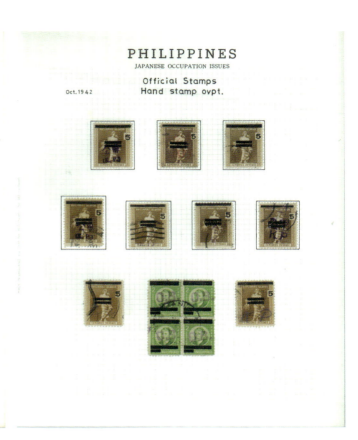

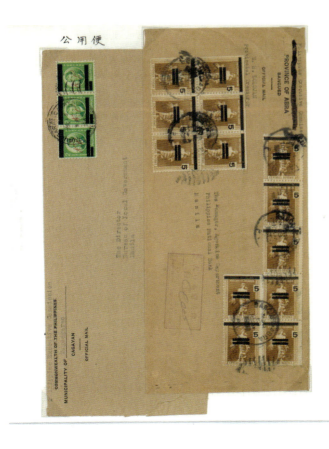

PHILIPPINES

Official cover
公用便

From Ormoc, Leyte.
to Tacloban, Leyte.
K.P. Red ovpt.

レイテ島 オルモック から
同島 タクロバン あて
赤ハン K.P. 加刷

9

PHILIPPINES

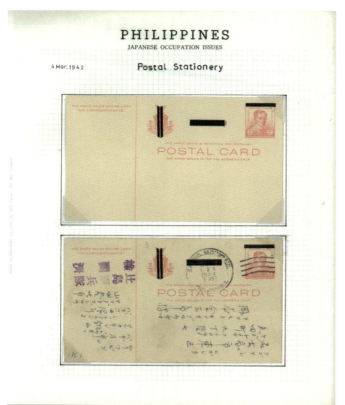

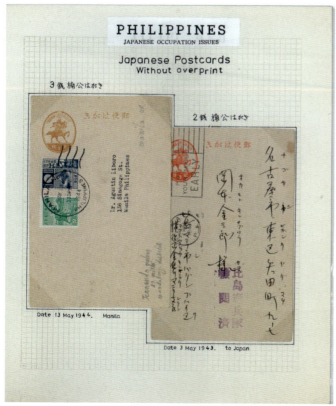

PHILIPPINES

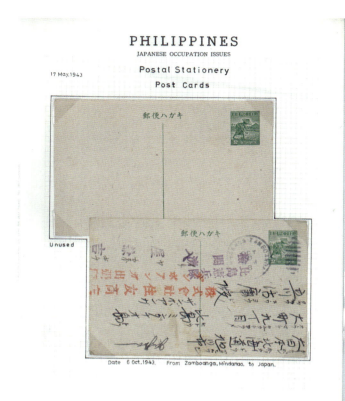

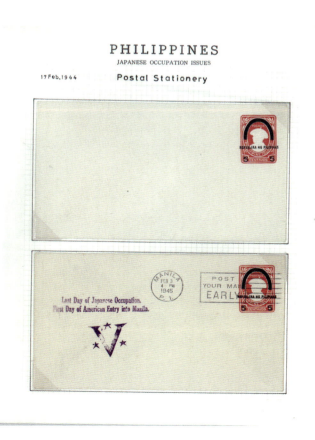

PHILIPPINES

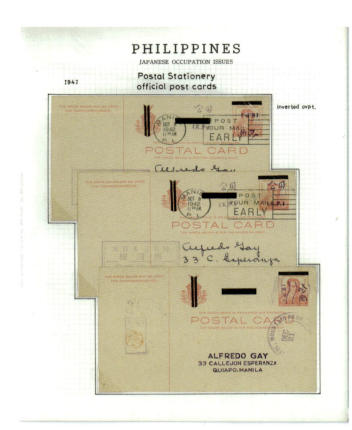

HONG KONG

1942年（昭和17年）1月　　　　　軍政庁廃止し総督部を設立
1942年（昭和17年）1月22日　　　日本軍郵政開始
1945年（昭和20年）4月16日　　　加刷切手発行
1945年（昭和20年）8月31日　　　日本軍郵政最終日

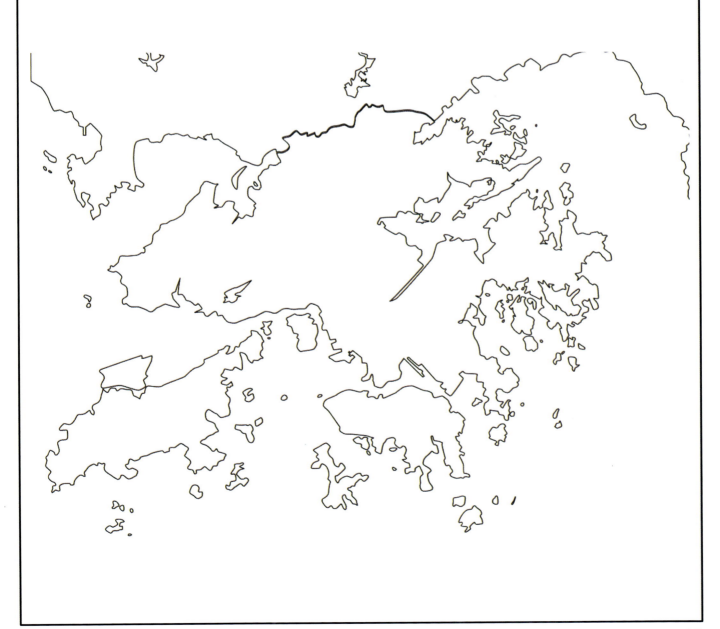

HONG KONG

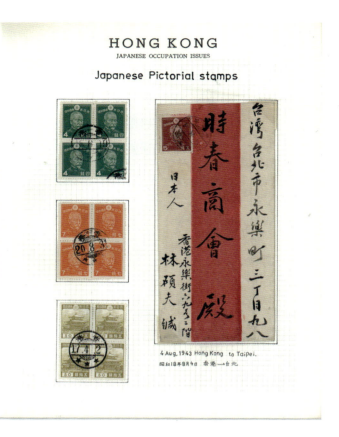

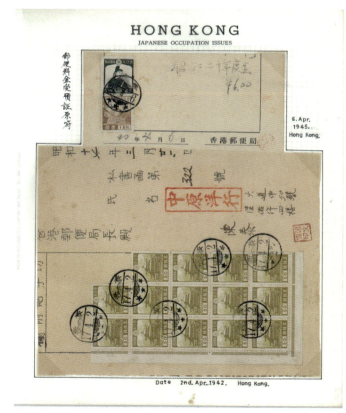

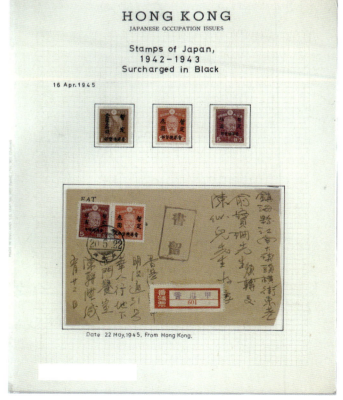

JAPANESE OCCUPATION "BALL FURONG" COLLECTION

HONG KONG

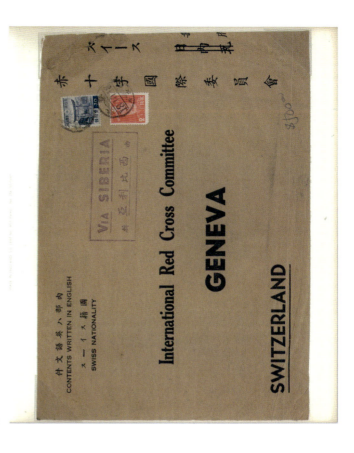

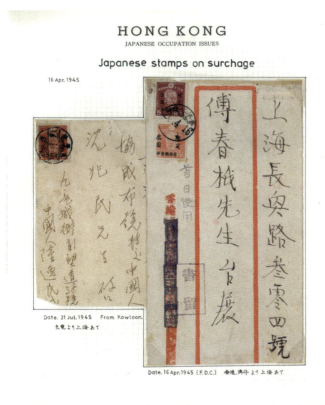

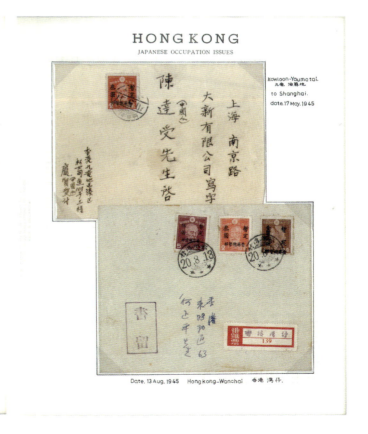

HONG KONG

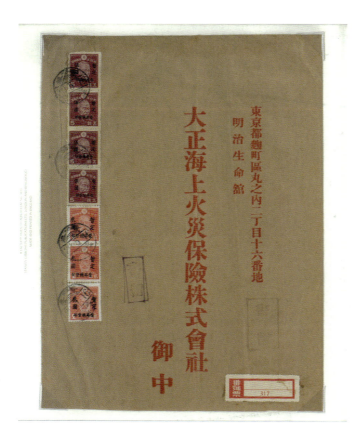

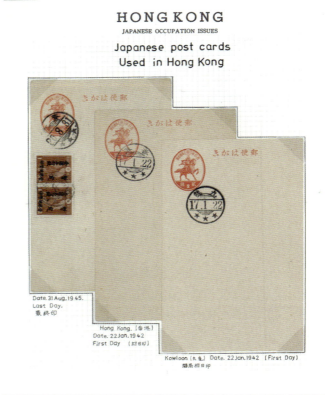

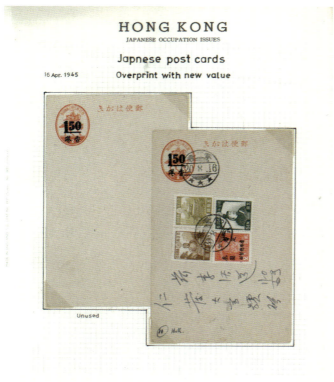

BURMA

1941年（昭和16年）12月14日	南部ビルマ　ビクトリアポイント　攻略
1942年（昭和17年）3月9日	ラングーン入城
1942年（昭和17年）3月15日	軍政部始動
1942年（昭和17年）5月1日	マンダレー占領
1942年（昭和17年）5月10日	治安維持委員会によるクジャク加刷切手発行
1942年（昭和17年）6月1日	矢野切手発行
1942年（昭和17年）6月7日	軍政を施行
1943年（昭和17年）8月1日	ビルマ独立、軍政を廃止
1945年（昭和20年）5月3日	英印軍　ラングーンに入城

BURMA

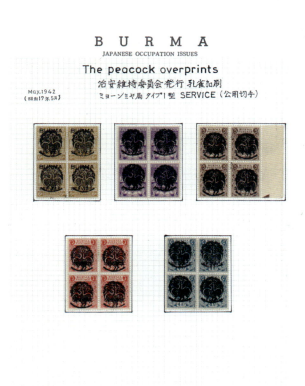
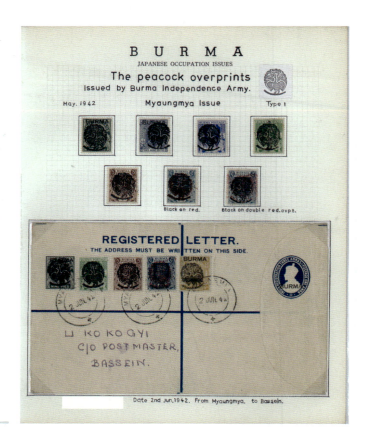
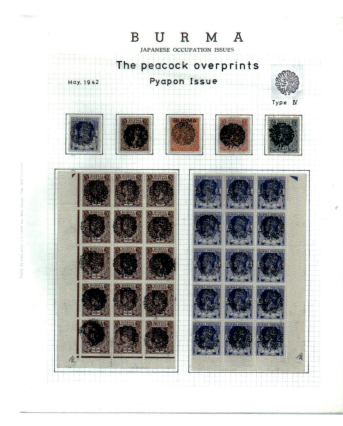
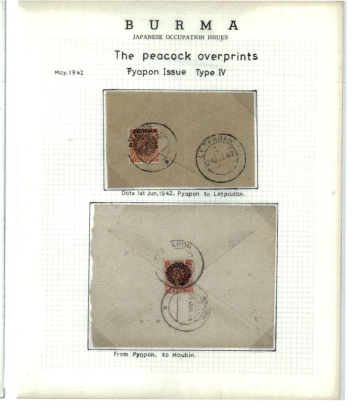

JAPANESE OCCUPATION "BALL FURONG" COLLECTION

BURMA

BURMA
JAPANESE OCCUPATION ISSUES
The peacock overprints
Myaungmya Issue Type I

May.1942 official stamps

Myaungmya Issue Type II

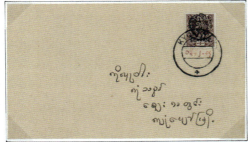

Date 22 May,1942 From Kyonpyaw.

BURMA
JAPANESE OCCUPATION ISSUES
The peacock overprints
Myaungmya Issue

Type II

Type III

BURMA
JAPANESE OCCUPATION ISSUES
The peacock overprints
Pyapon issue Type IV

May.1942. Sideways chop (機押抜エラー)

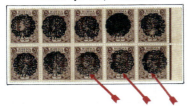

Henzada issue Type V

Post card Date 5th Jun.1942. From Henzada. to wakema.

BURMA
JAPANESE OCCUPATION ISSUES
The peacock overprints
Henzada Issue

May.1942

Type V

Double opt.

Date 1st Sep.1942 From Myanaung.

BURMA

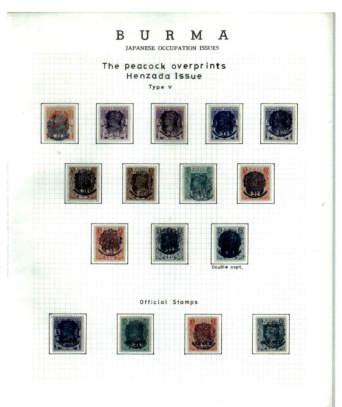

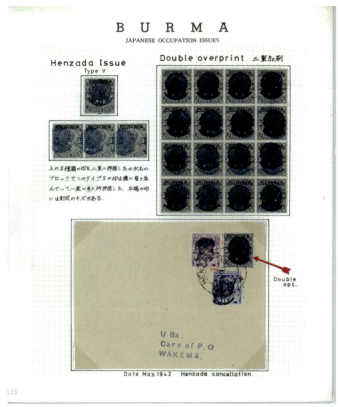

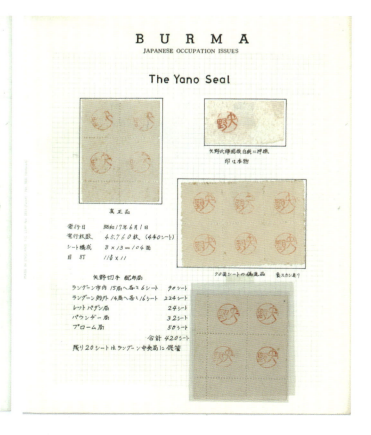

BURMA

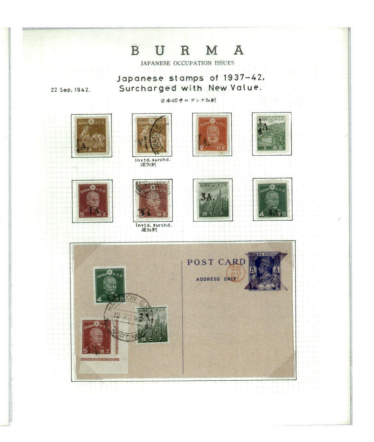

BURMA

JAPANESE OCCUPATION "BALL FURONG" COLLECTION

BURMA

BURMA
JAPANESE OCCUPATION ISSUES

First Regular Stamps Surcharged with New Value.
5 C. on 1 A.

15 Oct. 1942

BURMA
JAPANESE OCCUPATION ISSUES

Japanese stamps of 1937-42. Re-surcharged

日本切手にアンナ加刷 抹消、セント加刷

15 Oct. 1942

F.D.C.   昭日カバー 昭和17年10月15日 ラングーン

¼ A. invtd surch.

shan states

Date 15 Oct. 1942 (FDC)   From Bassein to Myaungmya.

Date 17 Oct. 1942

BURMA
JAPANESE OCCUPATION ISSUES

Japanese stamps of 1937-42, Surcharged in cents only.

日本切手にセント加刷

15 Oct. 1942

1C. omitted

shan states.

shan states.

23

南方占領地のフィラテリー　玉芙蓉コレクション

BURMA

JAPANESE OCCUPATION "BALL FURONG" COLLECTION

BURMA

BURMA

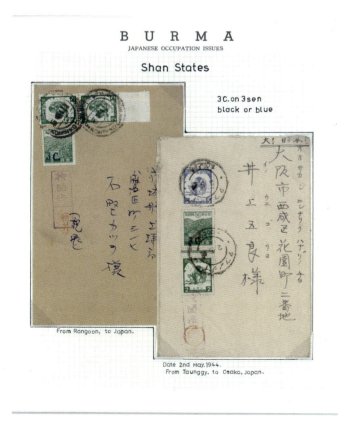

JAPANESE OCCUPATION "BALL FURONG" COLLECTION

BURMA

BURMA
JAPANESE OCCUPATION ISSUES

2nd Regular Stamps Used on covers

Date 3rd Jan. 1944

BURMA
JAPANESE OCCUPATION ISSUES

For Use in Shan States

1st Oct. 1943   シャン州切手

Imperf.

Colour-Proof

BURMA
JAPANESE OCCUPATION ISSUES

Overprinted Burmese

ビルマ文字加刷切手貼付カバー

Date 3rd Nov. 1944
From Rangoon
to Amarapura.

Date. 5th Feb. 1945. From Insein. to Rangoon.

BURMA
JAPANESE OCCUPATION ISSUES

Overprint inverted 2 Cent

NOV. 1944   逆加刷

2C. Inverted cover   2セント逆加刷切手貼付エンタイヤー
1944年11月14日 ラングーン消印

27

BURMA

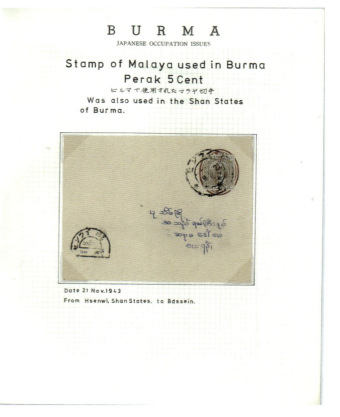

MALAYA

1941年（昭和16年）12月19日	ペナン島に上陸
1942年（昭和17年）1月31日	ケダー州アロースター局開局
1942年（昭和17年）2月15日	シンガポール占領
1942年（昭和17年）2月23日	軍政開始
1942年（昭和17年）2月25日	マライ郵政業務再開
1942年（昭和17年）3月7日	マライの行政機構を昭南特別市と10州に定める。
1942年（昭和17年）3月16日	昭南局（シンガポール）を開局　二重枠軍政印加刷発行
1943年（昭和18年）4月29日	正刷切手発行
1943年（昭和18年）10月19日	マライ北部4州をタイに割譲
1945年（昭和20年）9月3日	英軍　ペナンに上陸接収し、軍政を施行
1945年（昭和20年）9月5日	英軍　シンガポールに上陸接収し、軍政を施行

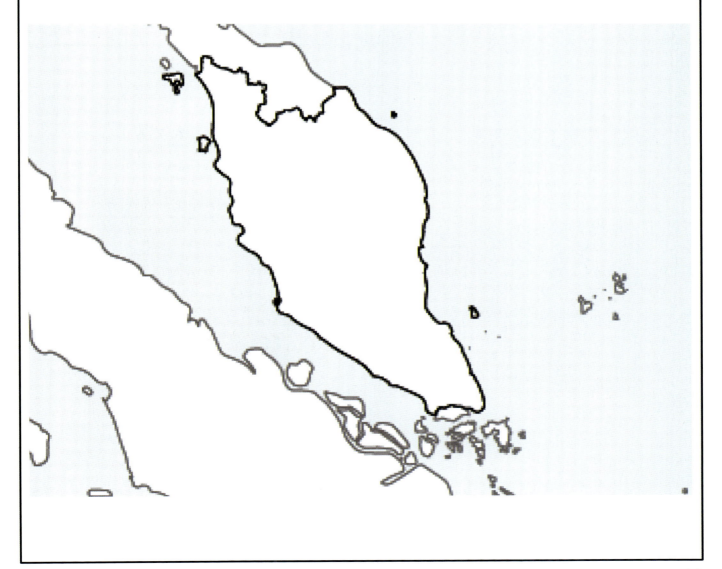

MALAYA

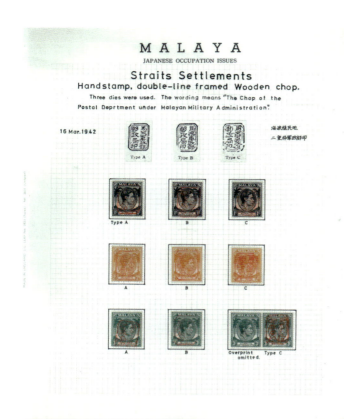

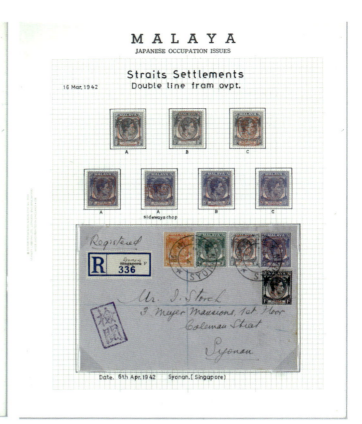

JAPANESE OCCUPATION "BALL FURONG" COLLECTION

MALAYA

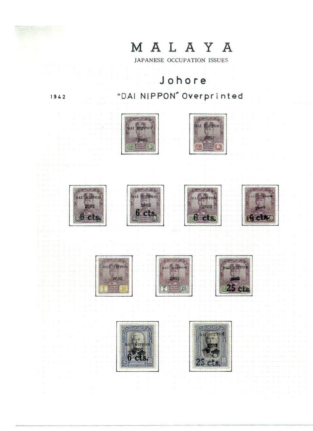

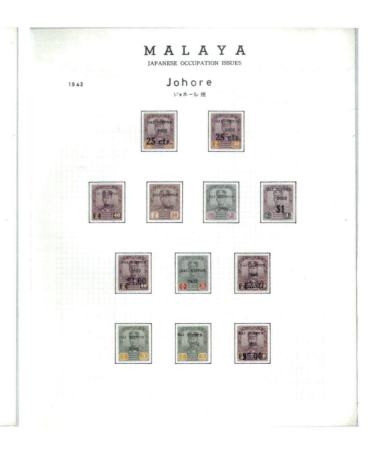

MALAYA

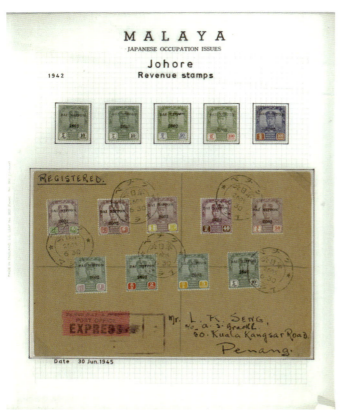

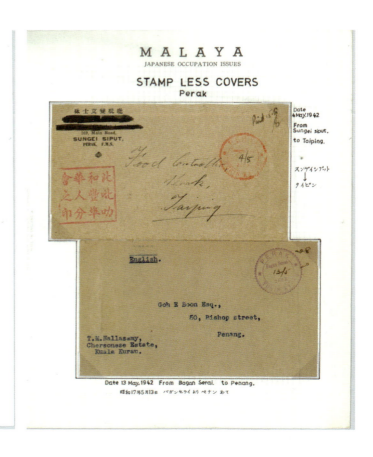

JAPANESE OCCUPATION "BALL FURONG" COLLECTION

MALAYA

MALAYA
JAPANESE OCCUPATION ISSUES

STAMP LESS COVER
Perak

ペラー州における初期無切手時期

Date 1st. Apr. 1942 to Ipoh, Perak.   First day cover.
昭和17年4月1日　ペラー州イポーあて　実逓初日カバー

At Perak, after being occupied by Japanese Army, Post Offices were reopened on 1st. Apr. 1942. As no stamps were available at opening time at first, letters or postcards were accepted at the counter of Post Office by writting date, charges and others on them by hand with signature. This provisional arrangement was put in practice two monthes from 1st. Apr. to 30th. May.

MALAYA
JAPANESE OCCUPATION ISSUES

STAMP LESS COVER
Perak

ペラー州の初期無切手カバー

From Ipoh. to Taiping.   Date 26 Apr. 1942
昭和17年4月26日　ペラー州イポー．より　同州タイピン あて．

MALAYA
JAPANESE OCCUPATION ISSUES

Penang
Okugawa Seal

奥川印

Mar. 1942

Okugawa Seal.   FDC. 30 Mar. 1942

MALAYA
JAPANESE OCCUPATION ISSUES

Penang
Okugawa Seal Type 1.

30 Mar. 1942

奥川印　大型

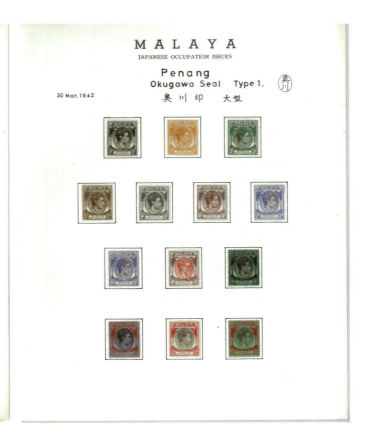

33

MALAYA

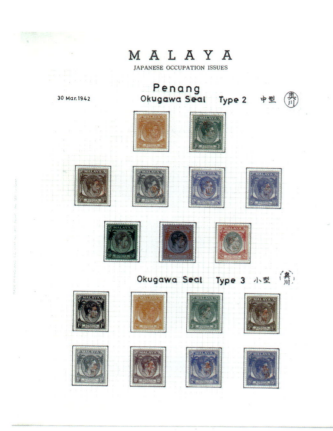

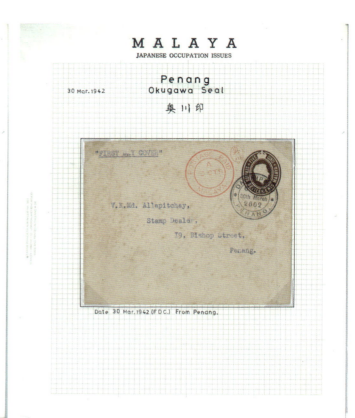

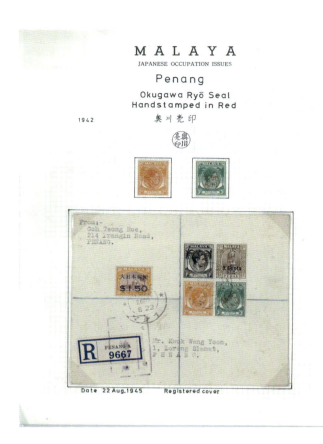
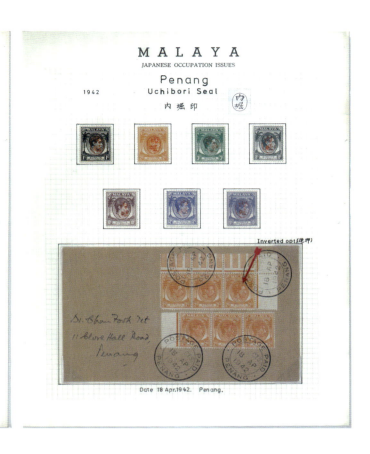

MALAYA

Date 7th Oct.1942 From Penang. to Kuala Kangsar, Perak.

Date 23 Apr.1942 (FDC) From Malacca. to Syonan (Singapore)

MALAYA

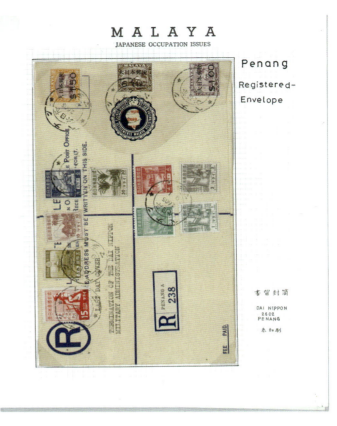

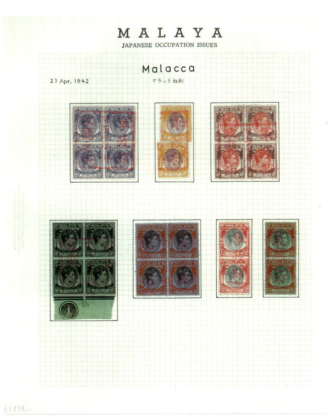

南方占領地のフィラテリー　玉芙蓉コレクション

MALAYA

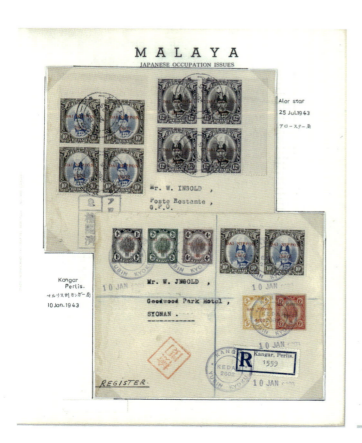

MALAYA

MALAYA

MALAYA
JAPANESE OCCUPATION ISSUES

Kedah

Date 13 May.1942 (FDC). Alor star.

Date 25 Jul.1942. From Alor star. to Singapor.

MALAYA
JAPANESE OCCUPATION ISSUES

Kedah
'DAI NIPPON 2602' Overprint

Post Card

Registered Envelope Date Sep.1942. From Baling, Kedah, to Syonan.
昭和17年9月 ケダー州バリン から 昭南あて

MALAYA
JAPANESE OCCUPATION ISSUES

Kelantan
Sunagawa Seal
砂川印

Date 9th Jul.1942. From Kota Bharu, Kelantan, to Penang.
昭和17年7月9日 ケランタン州 コタバル から ペナン あて

BACKSTAMP

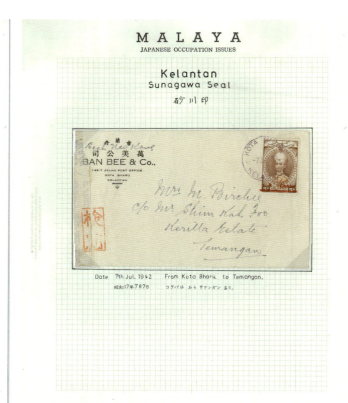

MALAYA
JAPANESE OCCUPATION ISSUES

Kelantan
Sunagawa Seal
砂川印

Date 7th Jul.1942. From Kota Bharu to Temangan.
昭和17年7月7日 コタバル から テマンガン あて

MALAYA

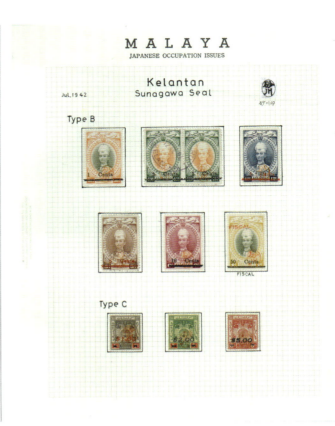

MALAYA

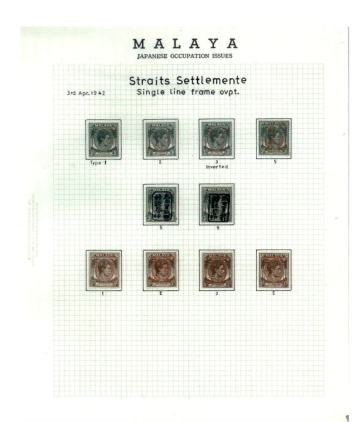

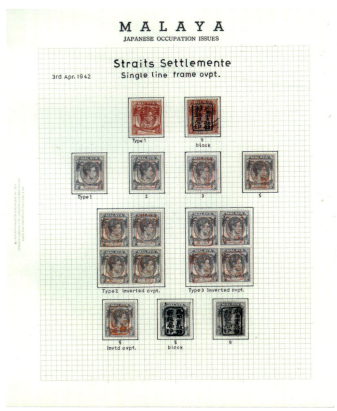

JAPANESE OCCUPATION "BALL FURONG" COLLECTION

MALAYA

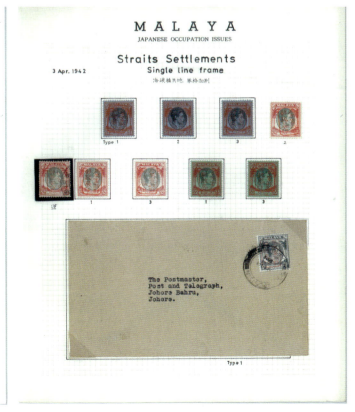

MALAYA

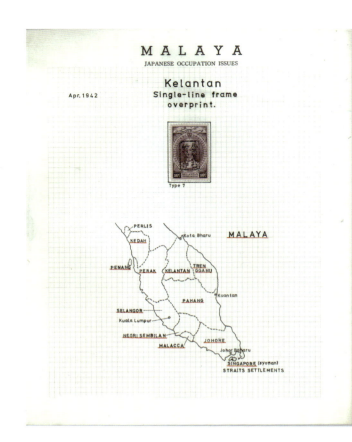

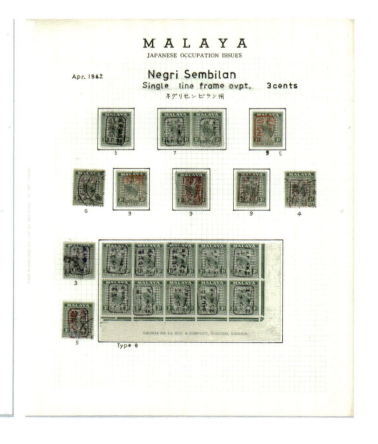

JAPANESE OCCUPATION "BALL FURONG" COLLECTION

MALAYA

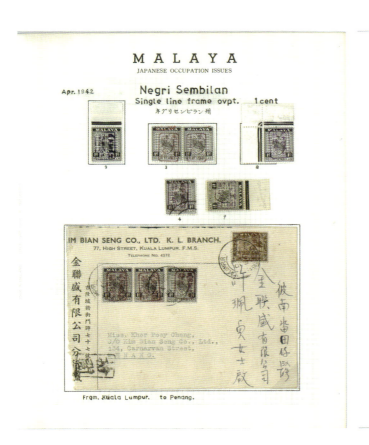
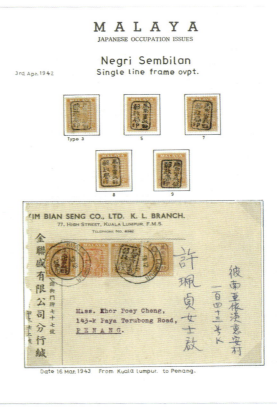

南方占領地のフィラテリー　玉芙蓉コレクション

MALAYA

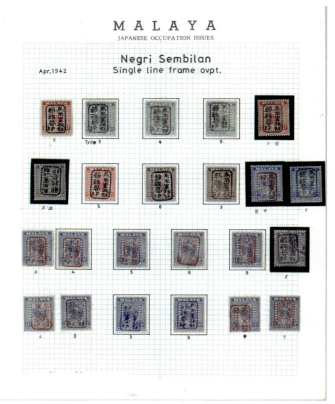

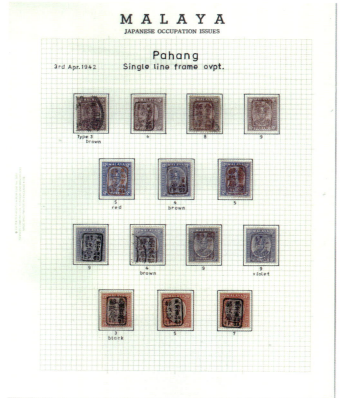
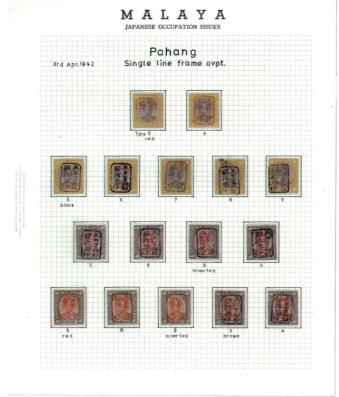

JAPANESE OCCUPATION "BALL FURONG" COLLECTION

MALAYA

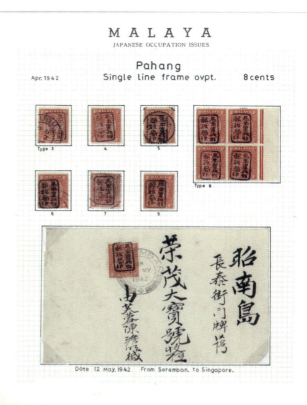
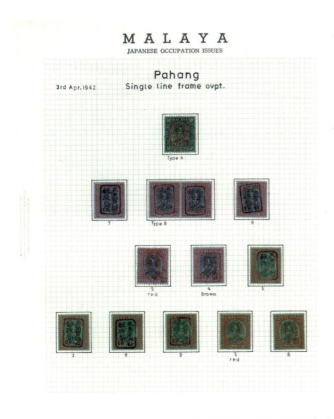

MALAYA

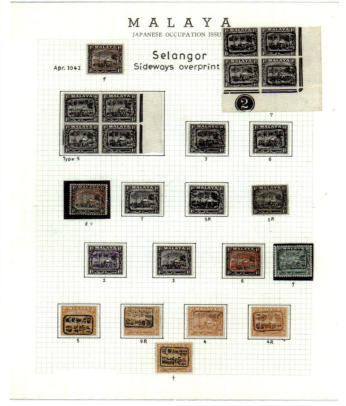

JAPANESE OCCUPATION "BALL FURONG" COLLECTION

MALAYA

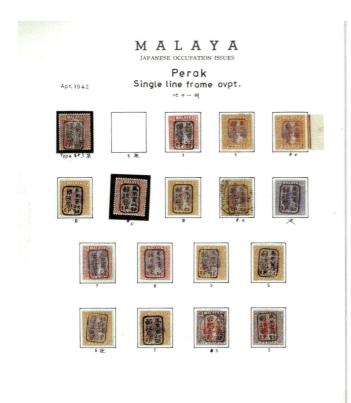

南方占領地のフィラテリー　玉芙蓉コレクション

MALAYA

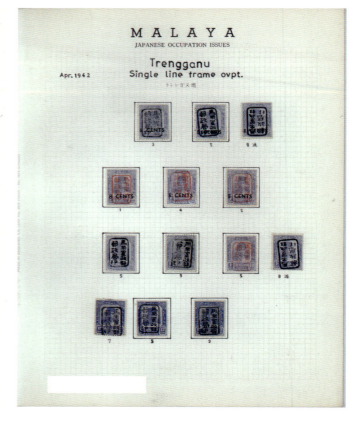

JAPANESE OCCUPATION "BALL FURONG" COLLECTION

MALAYA

MALAYA

Malaya — Japanese Occupation Issues — Straits Settlements — 3rd Apr. 1942 — Single line frame ovpt. Type 1

Date 10 June 1943. From Mersin. to Syonan. Johre.

Malaya — Japanese Occupation Issues — Straits Settlements — 3rd. Apr. 1942 — Single line frame ovpt. Type 2

Date 6th. June 1942. From Syonan. to Kuala Kangsar.

MALAYA

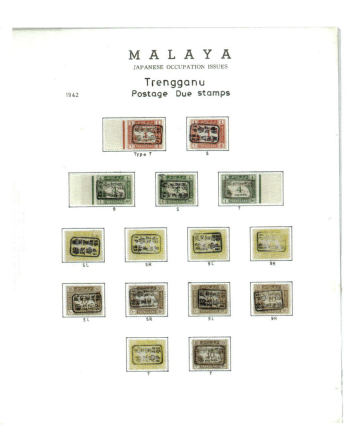

MALAYA

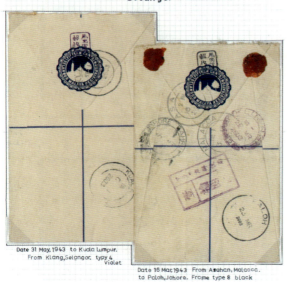

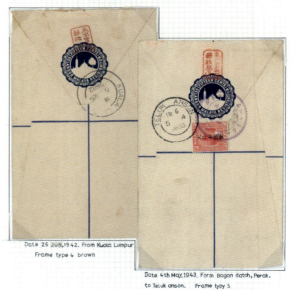

MALAYA

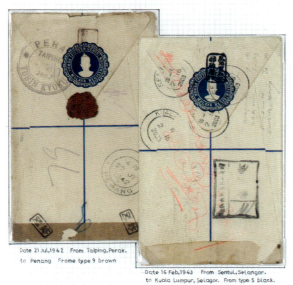

Straits Settlements
May.1942 — 'DAI NIPPON' overprinted in blak

Negri Sembilan
Jun.1942 — "DAI NIPPON 2602 MALAYA" Ovpt.

Inverted ovpt.   Double one invtd.

MALAYA

JAPANESE OCCUPATION "BALL FURONG" COLLECTION

MALAYA

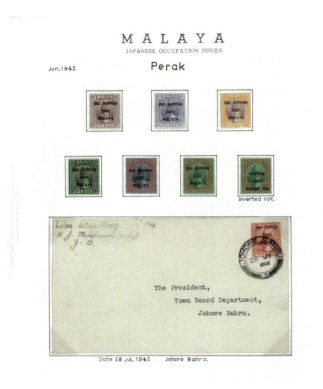

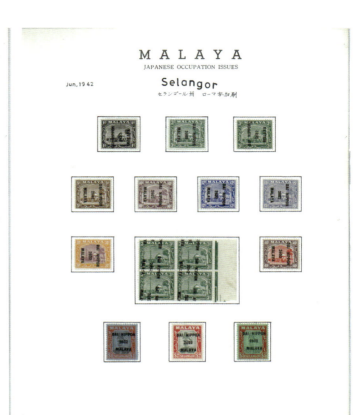

MALAYA

JAPANESE OCCUPATION "BALL FURONG" COLLECTION

MALAYA

MALAYA
JAPANESE OCCUPATION ISSUES

Post Cards
Pahang

Date 21 Jul, 1942. From Ipoh, Perak, to Taiping, Perak.

MALAYA
JAPANESE OCCUPATION ISSUES

Post Cards
Perak

Date 6th Nov, 1942. From Kuala Lumpur, to Kangsar.

MALAYA
JAPANESE OCCUPATION ISSUES

"DAI NIPPON YUBIN" Overprint
Perak

Nov. 1942

Wide space between '2' & 'Cents'

inverted overprint

'2 Cents' omitted

Selangor

MALAYA
JAPANESE OCCUPATION ISSUES

Post Cards
Perak
'DAI NIPPON YUBIN' ovpt.

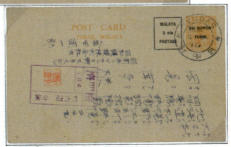

Date 5th Dec, 1942. From Syonan, to Japan.

MALAYA

JAPANESE OCCUPATION "BALL FURONG" COLLECTION

MALAYA

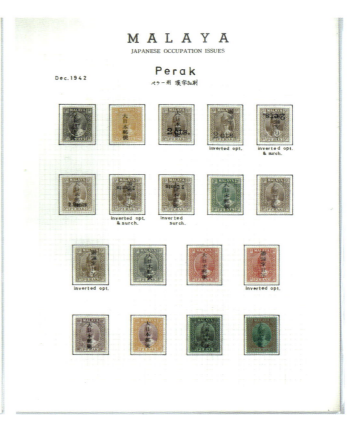

MALAYA

MALAYA
JAPANESE OCCUPATION ISSUES
Post Cards
Straits Settlements — Gleen print

Date 31 Dec, 1944. Form Penang. to Osako, Japan.

MALAYA
JAPANESE OCCUPATION ISSUES
Post Cards
Straits Settlements — Orange print

Date 5th Jul, 1943. From Syonan. to Kuala Kangsar, Perak.

MALAYA
JAPANESE OCCUPATION ISSUES
Post Cards
Trengganu

Date 19 Jul, 1943. From Syonan. to Kuala Kangsar.

MALAYA
JAPANESE OCCUPATION ISSUES
Registered Envelopes
Straits Settements
Black overprint

Date 5 spt. 1943 From Pontian, Johore. to Johore Bahru.

MALAYA

MALAYA
JAPANESE OCCUPATION ISSUES

Post Cards — 2 cts (small 'c')
Perak

MALAYA
JAPANESE OCCUPATION ISSUES

Post Cards — 2 Cts (large 'c')
Perak

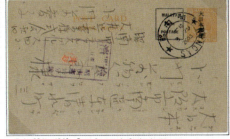

Date 29 Mar, 1943. From Syonan to Osaka, Japan.

Date 29 June, 1943. From Syonan to Syonan.

MALAYA
JAPANESE OCCUPATION ISSUES

Registered Envelopes
Johore

Date 7th Apr, 1944. From Syonan (Singapore)

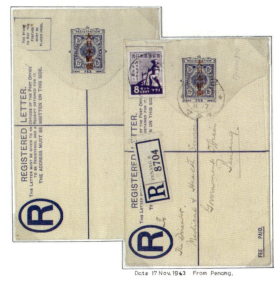

MALAYA
JAPANESE OCCUPATION ISSUES

Registered Envelopes
Kedah

Date 17 Nov, 1943. From Penang.

MALAYA

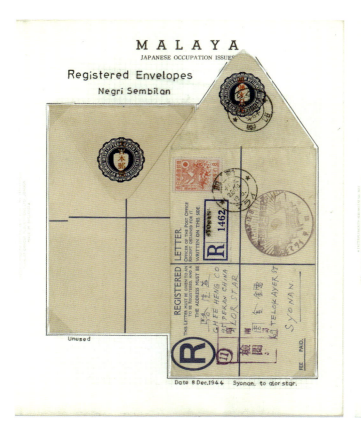

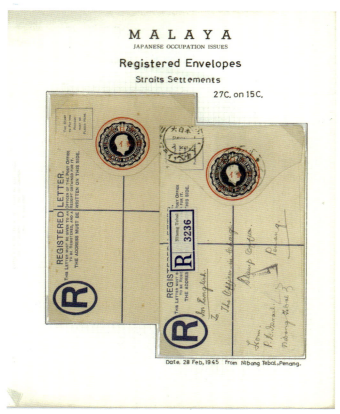

JAPANESE OCCUPATION "BALL FURONG" COLLECTION

MALAYA

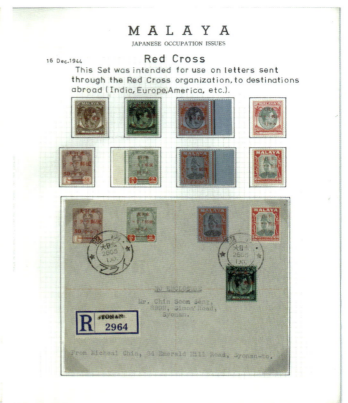

MALAYA

JAPANESE OCCUPATION "BALL FURONG" COLLECTION

MALAYA

67

MALAYA

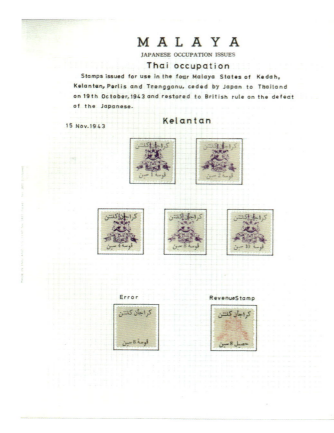

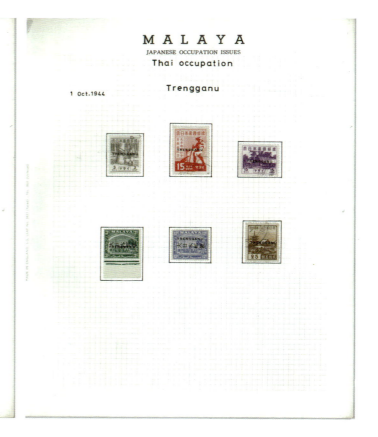

NORTH BORNEO

1941年（昭和16年）12月16日	ミリを占領
	ミリ・クチンの要衝を占領した直後に軍政本部を設立、12月23日には無加刷のまま継続使用。
1942年（昭和17年）4月	軍政の滲透
1942年（昭和17年）7月15日	西ボルネオを陸軍から海軍へ
1942年（昭和17年）10月1日	加刷切手発行
1943年（昭和18年）4月29日	正刷切手発行
1945年（昭和20年）9月4日	郵便局の閉鎖

NORTH BORNEO

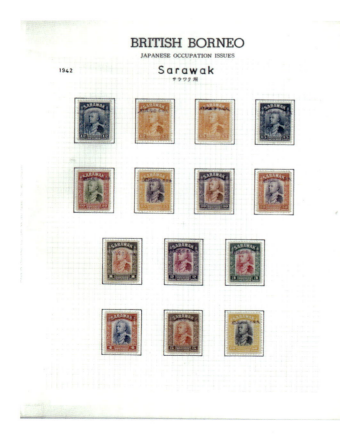

JAPANESE OCCUPATION "BALL FURONG" COLLECTION

NORTH BORNEO

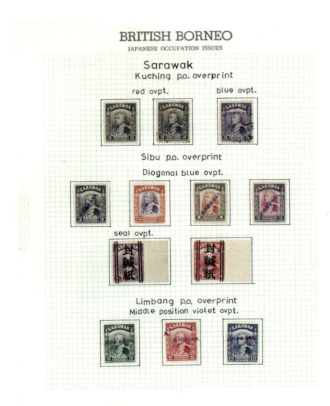

NORTH BORNEO

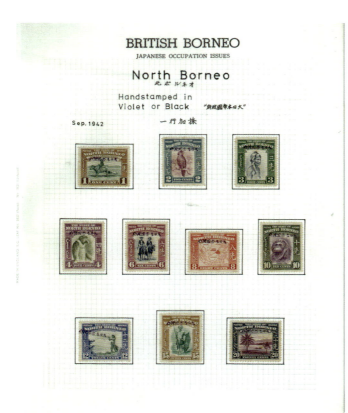
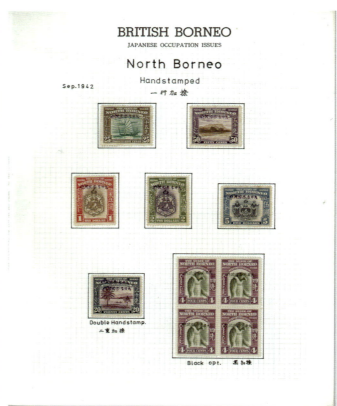

JAPANESE OCCUPATION "BALL FURONG" COLLECTION

NORTH BORNEO

BRITISH BORNEO
JAPANESE OCCUPATION ISSUES

Oct. 1942 **North Borneo**

WAR TAX   Postage stamp Currency

Date 2nd Oct.1944 Kuching, Registered cover

BRITISH BORNEO
JAPANESE OCCUPATION ISSUES

Postage Due stamps
North Borneo

Stamps for seal ovpt.

BRITISH BORNEO
JAPANESE OCCUPATION ISSUES

Oct.1942. **Brunei** One line Japanese handstamped ovpt.

Violet

Blue

Red

BRITISH BORNEO
JAPANESE OCCUPATION ISSUES

Brunei

Date 31 Aug 1943
From Kuching
to Osaka, Japan.
Registered cover

73

NORTH BORNEO

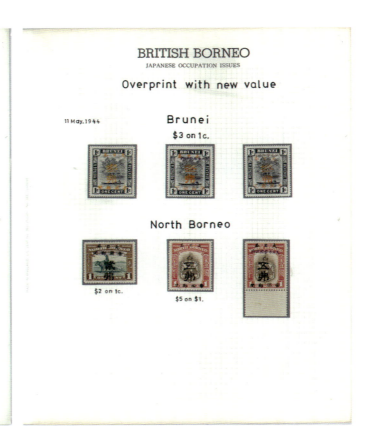

JAPANESE OCCUPATION "BALL FURONG" COLLECTION

NORTH BORNEO

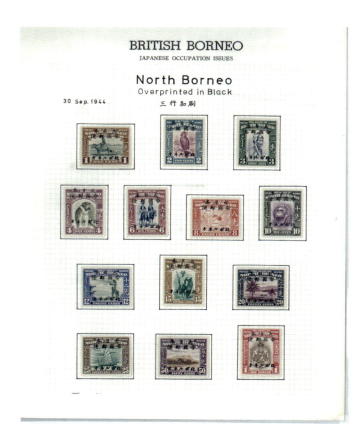

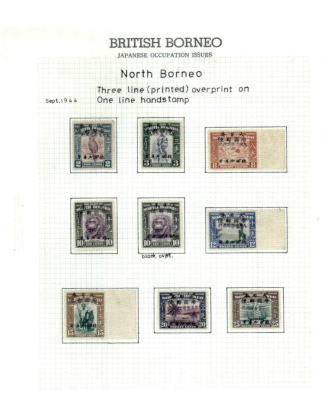

NORTH BORNEO

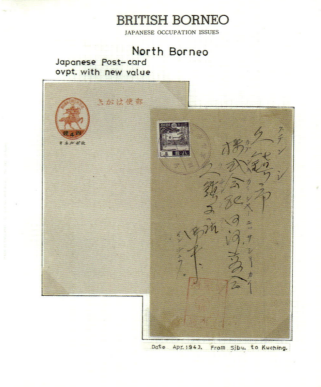
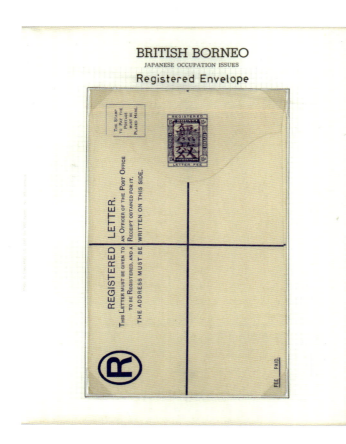
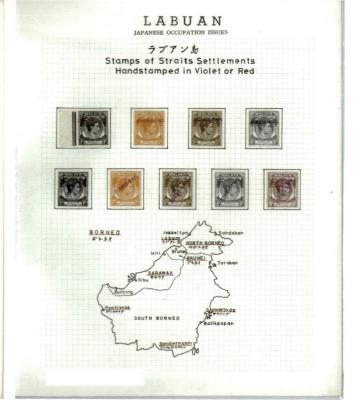

SUMATRA

1942年（昭和17年）2月15日　　　パレンバン占領
1942年（昭和17年）2月23日　　　軍政開始
1942年（昭和17年）3月 中旬　　　メダン占領
1942年（昭和17年）3月18日　　　クタラジャ局（アチェ州）を開局
1942年（昭和17年）3月20日　　　スマトラに軍政を実施
1942年（昭和17年）3月24日　　　パレンバン州各局を開局　パレンバンローカル加刷発行
1943年（昭和18年）4月29日　　　正刷切手発行
1943年（昭和18年）5月1日　　　　マライから分離ブキチンギに軍政監部を置く
1945年（昭和20年）8月17日　　　インドネシア独立宣言

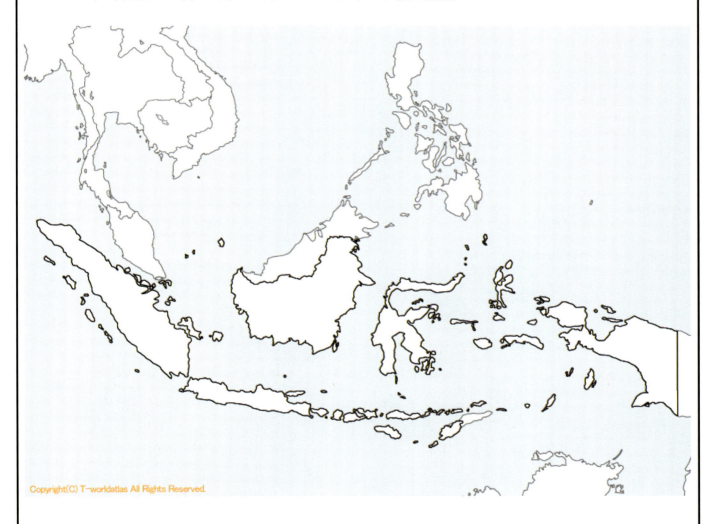

SUMATRA

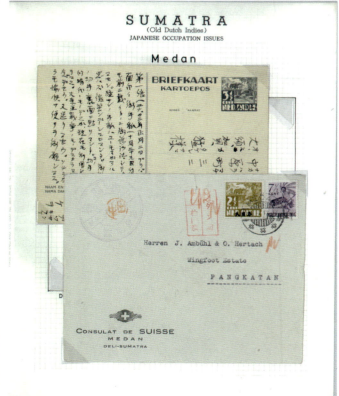

JAPANESE OCCUPATION "BALL FURONG" COLLECTION

SUMATRA

79

南方占領地のフィラテリー　玉芙蓉コレクション

SUMATRA

JAPANESE OCCUPATION "BALL FURONG" COLLECTION

SUMATRA

81

SUMATRA

SUMATRA

SUMATRA
(Old Dutch Indies)
JAPANESE OCCUPATION ISSUES

Sumatra's East Coast
Medan
Dainippn in frame Red.
Dancer series

Feb. 1943

inverted ovpt.

13 Jan. 1944

SUMATRA
(Old Dutch Indies)
JAPANESE OCCUPATION ISSUES

Sumatra's East Coast
Dainippon in frame Red overprint

Date 10 Apr. 1943. From Pematangsiantar. to Limapoeloeh.

SUMATRA
(Old Dutch Indies)
JAPANESE OCCUPATION ISSUES

Achin
Koeta Radja overprint
Type 1 (fine)
Postage Due stamps
Enschede

1942

Kolff

SUMATRA
(Old Dutch Indies)
JAPANESE OCCUPATION ISSUES

Achin
Koeta Radja overprint
Type 2 (coarse)
UNWMKD

1942

Date 3rd Sep. 1942. From Sigli. to Medan.

南方占領地のフィラテリー　玉芙蓉コレクション

SUMATRA

SUMATRA

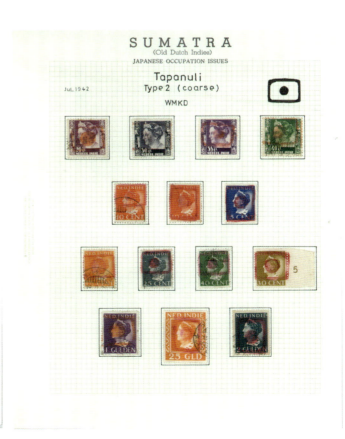

SUMATRA

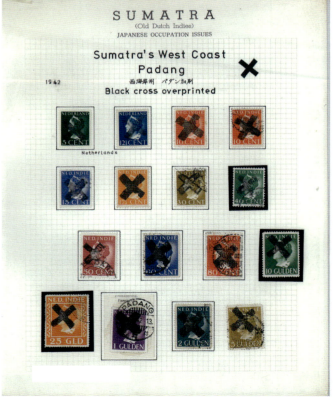

SUMATRA

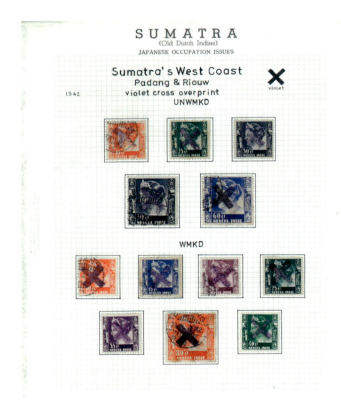
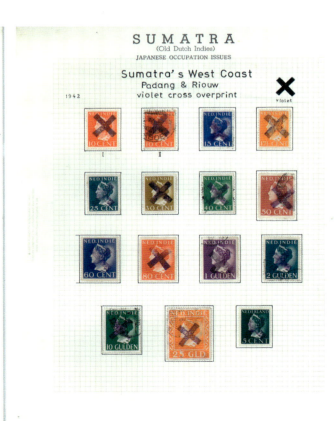

SUMATRA

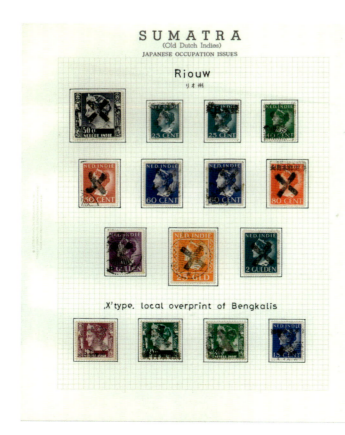

SUMATRA

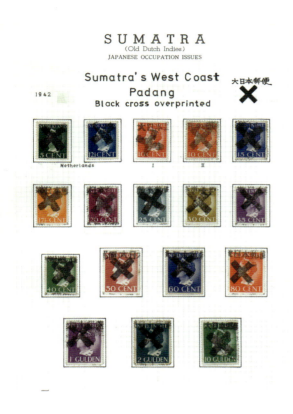

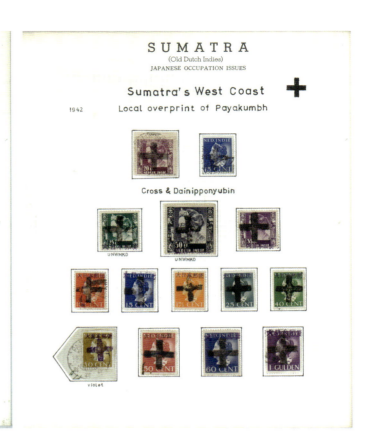

SUMATRA

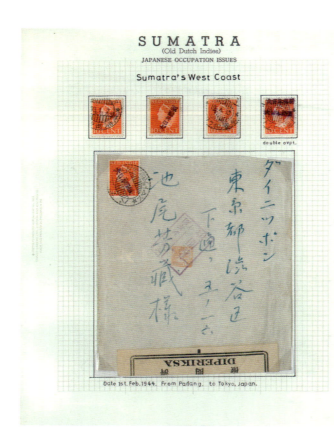

SUMATRA

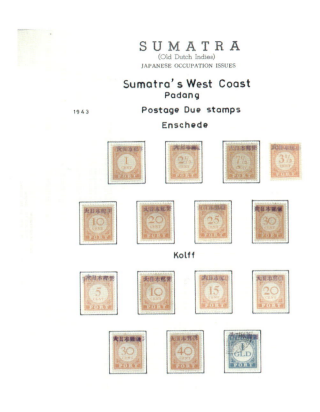

SUMATRA

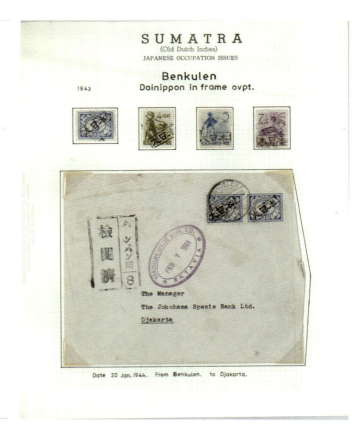

JAPANESE OCCUPATION "BALL FURONG" COLLECTION

SUMATRA

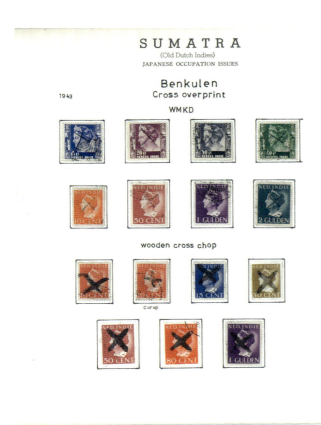

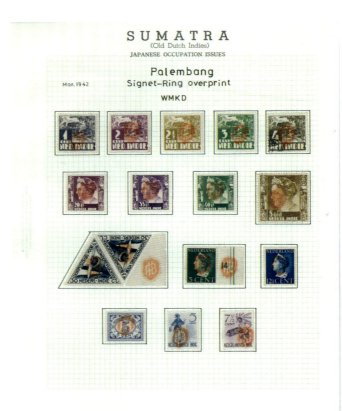

南方占領地のフィラテリー　玉芙蓉コレクション

SUMATRA

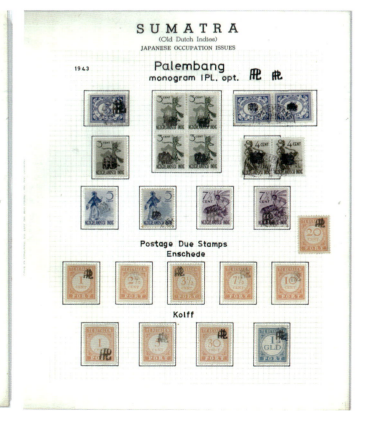

JAPANESE OCCUPATION "BALL FURONG" COLLECTION

SUMATRA

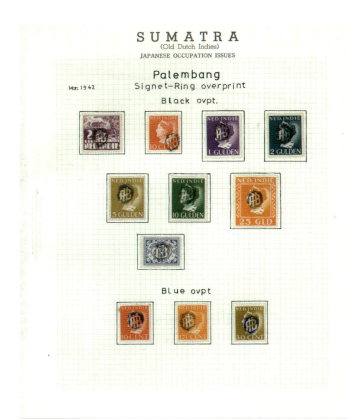

SUMATRA

JAPANESE OCCUPATION "BALL FURONG" COLLECTION

SUMATRA

SUMATRA

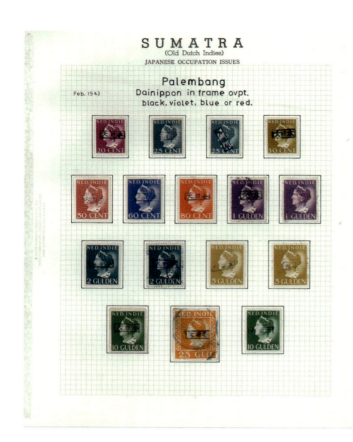

JAPANESE OCCUPATION "BALL FURONG" COLLECTION

SUMATRA

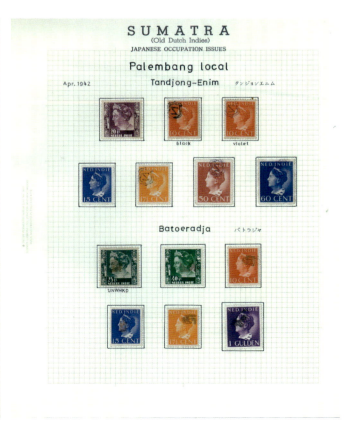

99

SUMATRA

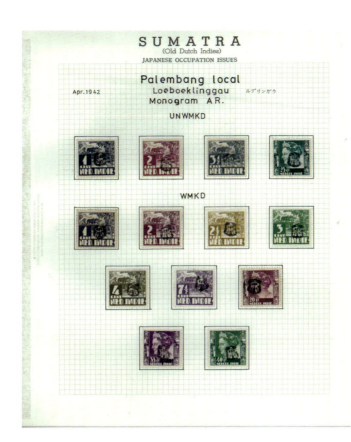

JAPANESE OCCUPATION "BALL FURONG" COLLECTION

SUMATRA

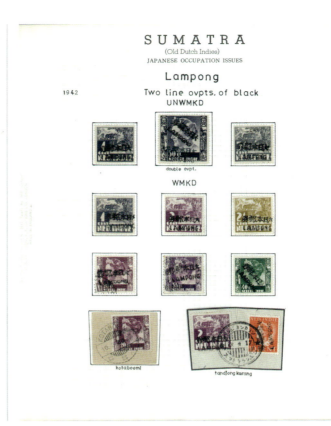

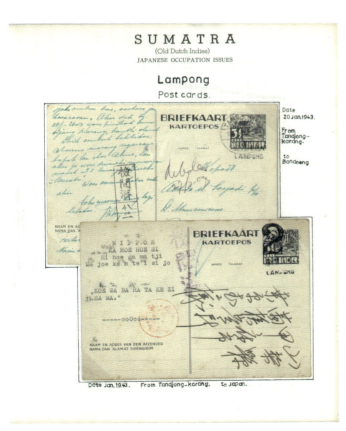

SUMATRA

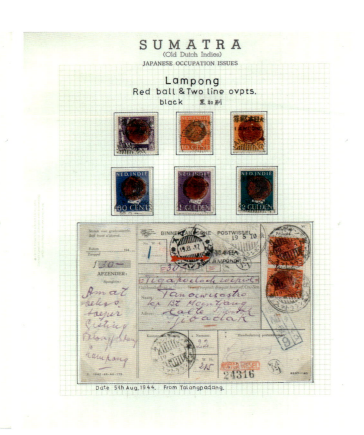

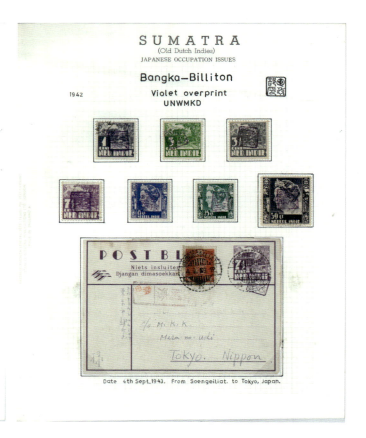

JAPANESE OCCUPATION "BALL FURONG" COLLECTION

SUMATRA

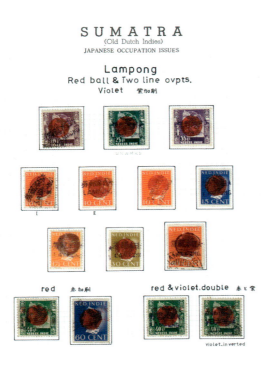

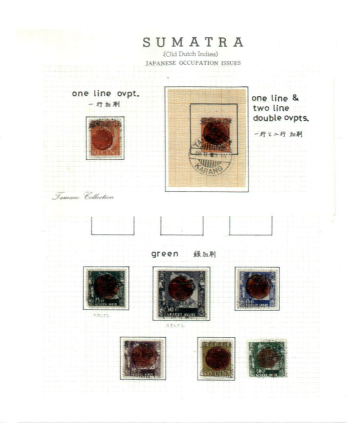

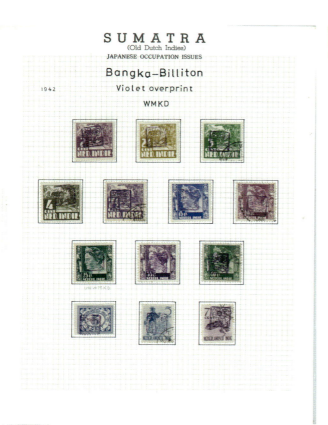

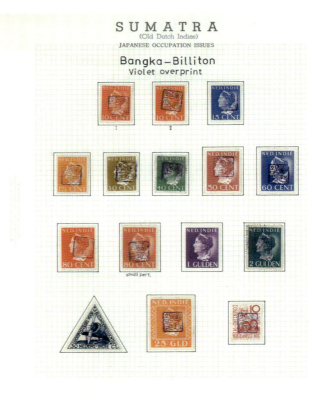

SUMATRA

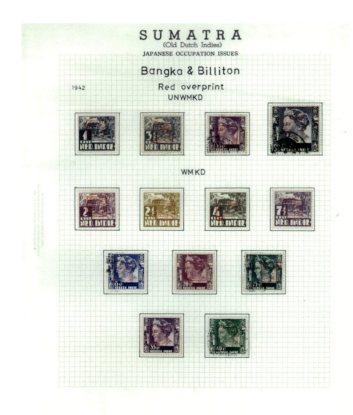

SUMATRA

SUMATRA

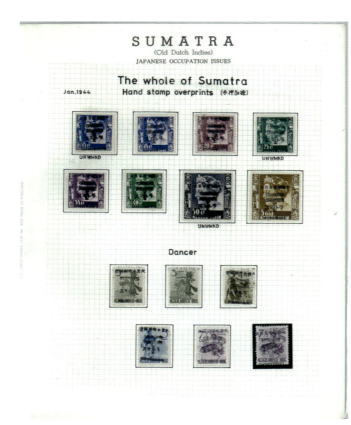

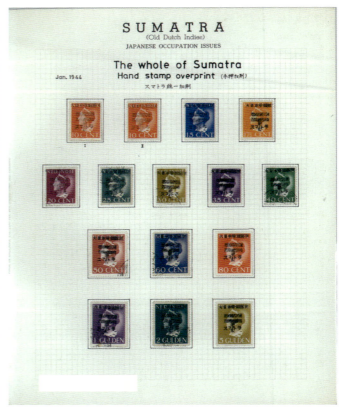

SUMATRA

SUMATRA

JAPANESE OCCUPATION "BALL FURONG" COLLECTION

SUMATRA

SUMATRA

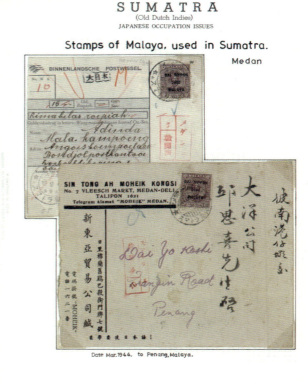

JAPANESE OCCUPATION "BALL FURONG" COLLECTION

SUMATRA

SUMATRA
(Old Dutch Indies)
JAPANESE OCCUPATION ISSUES

29 Apr. 1943   Regular post cards.

Date 3rd. Oct. 1944. From Manggar, Billiton, to Japan.

SUMATRA
(Old Dutch Indies)
JAPANESE OCCUPATION ISSUES

Unoverprinted

On covers (無加刷切手)

Date 13 Aug.1942 Bindjei.

Date 23 Aug.1942 Tebingtinggi-deli. to Medan.

SUMATRA
JAPANESE OCCUPATION ISSUES

SUMATRA
(Old Dutch Indies)
JAPANESE OCCUPATION ISSUES

Japanese Stamps Used in Sumatra

日本切手のスマトラ使用

Date. 19 Dec.1943.   From Medan. to Pematang Siantar.

SUMATRA

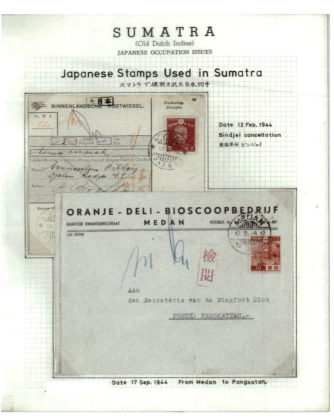

JAVA

1942年（昭和17年）3月9日	占領　軍政の布告
1942年（昭和17年）3月23日	バンドン郵便局再開
	地域によっては継続して業務を行っている
1942年（昭和17年）8月	正刷はがきの発行
1943年（昭和18年）3月9日	正刷記念切手の発行
1943年（昭和18年）4月29日	正刷普通切手の発行
1945年（昭和20年）8月17日	インドネシア独立宣言

JAVA

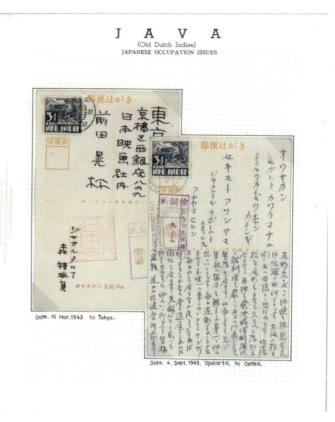

JAPANESE OCCUPATION "BALL FURONG" COLLECTION

JAVA

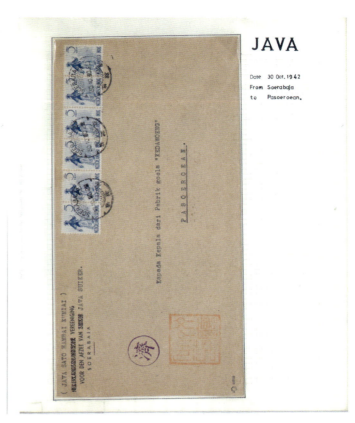

JAVA

JAPANESE OCCUPATION "BALL FURONG" COLLECTION

JAVA

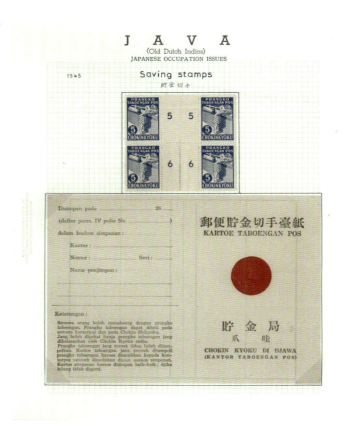

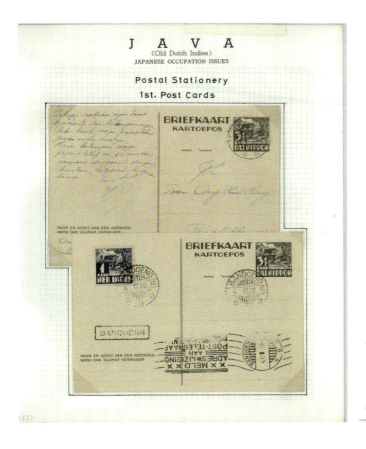

117

JAVA

JAPANESE OCCUPATION "BALL FURONG" COLLECTION

JAVA

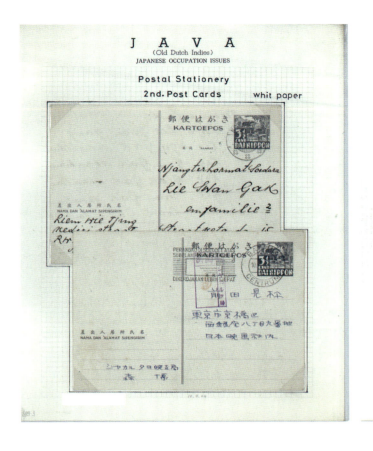

Date 21 Feb. 1943. From Bentjoeloek, to Soerabaja.

JAVA

NAVAL CONTROL AREA

東南ボルネオ
1942 年（昭和 17 年）1 月 24 日　　　　バリクパパン占領
1942 年（昭和 17 年）2 月 10 日　　　　バンジェルマシン占領
1942 年（昭和 17 年）3 月 1 日　　　　郵便局開局
1942 年（昭和 17 年）6 月 15 日　　　　錨加刷切手の登場

西ボルネオ
1942 年（昭和 17 年）7 月 15 日　　　　海軍による民政が始まる
1942 年（昭和 17 年）9 月　　　　　　　錨加刷切手の登場

セレベス
1942 年（昭和 17 年）6 月 20 日　　　　マカッサル開局、セレベス民生部正刷葉書発行
1942 年（昭和 17 年）7 月　　　　　　　錨加刷切手の登場

モルッカ諸島
1942 年（昭和 17 年）1 月 30 日　　　　アンボン上陸
1942 年（昭和 17 年）12 月　　　　　　加刷葉書の登場
1943 年（昭和 18 年）2 月 1 日　　　　フローレス暫定切手の発行

以上全域
1943 年（昭和 18 年）7 月 2 日　　　　正刷切手発行
1945 年（昭和 20 年）8 月 17 日　　　　インドネシア独立宣言
1945 年（昭和 20 年）9 月 15 日　　　　連合軍バンジェルマシンに

アンダマン
1942 年（昭和 17 年）3 月 20 日　　　　占領
1943 年（昭和 18 年）4 月 5 日　　　　加刷切手発行
1945 年（昭和 20 年）9 月　　　　　　　イギリス軍による再占領

NAVAL CONTROL AREA

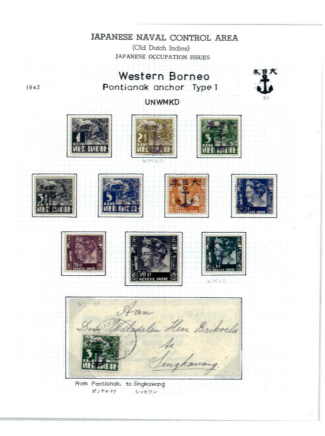

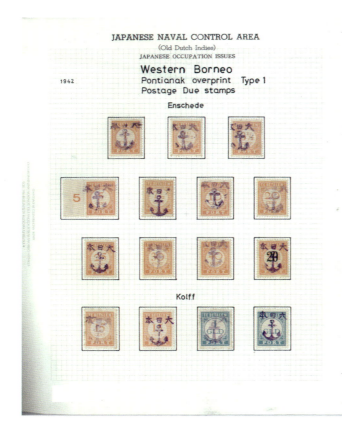

JAPANESE OCCUPATION "BALL FURONG" COLLECTION

NAVAL CONTROL AREA

JAPANESE NAVAL CONTROL AREA
(Old Dutch Indies)
JAPANESE OCCUPATION ISSUES

Western Borneo
Pontianak overprint
Type 1

1942

JAPANESE NAVAL CONTROL AREA
(Old Dutch Indies)
JAPANESE OCCUPATION ISSUES

Western Borneo
Pontianak anchor Type 1

red lilac

Post Card   Date 26 Jun.1943 From Pontianak. to Bandoeng.

JAPANESE NAVAL CONTROL AREA
(Old Dutch Indies)
JAPANESE OCCUPATION ISSUES

Western Borneo
Pontianak anchor Type 1
Red overprint

Date 8th Jun.1944. Pamangkat. to Pontianak.

Western Borneo
Pontianak
Date
5th Apr. 1943

123

NAVAL CONTROL AREA

JAPANESE OCCUPATION "BALL FURONG" COLLECTION

NAVAL CONTROL AREA

125

NAVAL CONTROL AREA

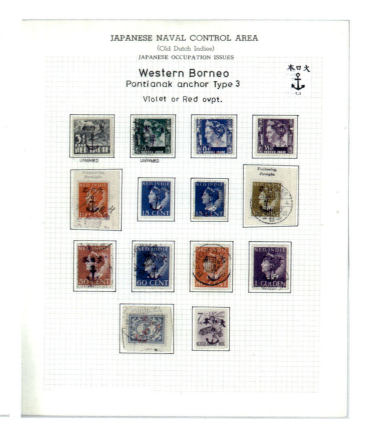

JAPANESE OCCUPATION "BALL FURONG" COLLECTION

NAVAL CONTROL AREA

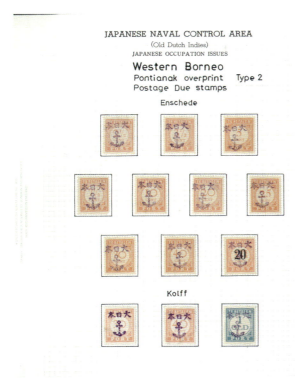

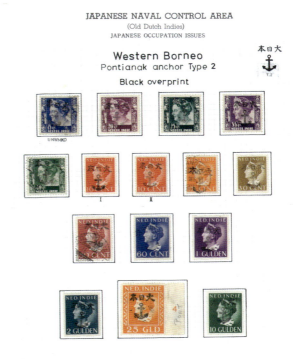

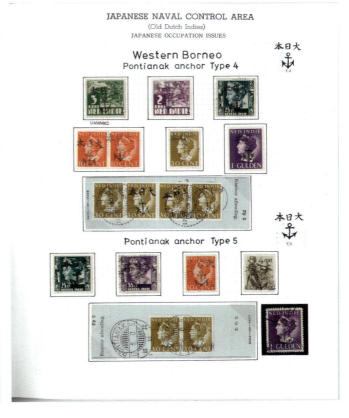

南方占領地のフィラテリー　玉芙蓉コレクション

NAVAL CONTROL AREA

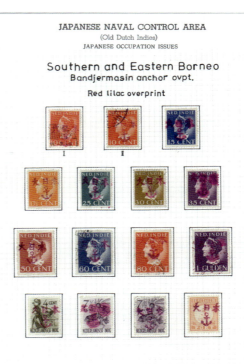

128

JAPANESE OCCUPATION "BALL FURONG" COLLECTION

NAVAL CONTROL AREA

JAPANESE NAVAL CONTROL AREA
(Old Dutch Indies)
JAPANESE OCCUPATION ISSUES

Southern and Eastern Borneo
Bandjermasin anchor ovpt.
Violet overprint
UNWMKD

WMKD

Newspaper wrapper

Date 17 May. 1943. From Balikpapan. to Pontianak.

JAPANESE NAVAL CONTROL AREA
(Old Dutch Indies)
JAPANESE OCCUPATION ISSUES

Southern and Eastern Borneo
1942 Bandjermasin overprint
Violet overprint

Date 4 Feb.1944 Sangkoelirang.

JAPANESE NAVAL CONTROL AREA
(Old Dutch Indies)
JAPANESE OCCUPATION ISSUES

Balikpapan

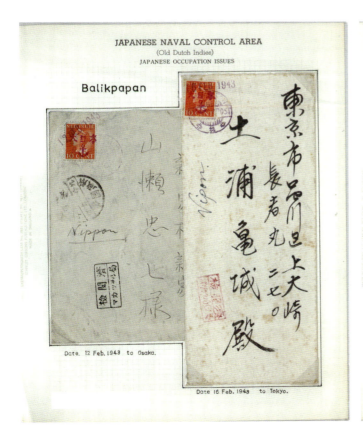

Date. 12 Feb. 1943 to Osaka.
Date 16 Feb. 1943 to Tokyo.

JAPANES NAVAL CONTROL AREA
(Old Dutch Indies)
JAPANESE OCCUPATION ISSUES

Bandjermasin

Date 12 Mar. 1943 From Bandjermasin. to Balik papan.

129

南方占領地のフィラテリー　玉芙蓉コレクション
NAVAL CONTROL AREA

JAPANESE OCCUPATION "BALL FURONG" COLLECTION

NAVAL CONTROL AREA

JAPANESE NAVAL CONTROL AREA
(Old Dutch Indies)
JAPANESE OCCUPATION ISSUES

Bandjermasin
Provisional issue

1944

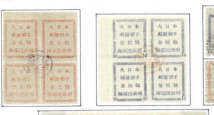

USED

JAPANESE NAVAL CONTROL AREA
(Old Dutch Indies)
JAPANESE OCCUPATION ISSUES

Samarinda
1945 サマリンダ加刷

Date 1st Feb. 1945. From Samarinda. to Djokjakarta.

JAPANESE NAVAL CONTROL AREA
(Old Dutch Indies)
JAPANESE OCCUPATION ISSUES

South Celebes
Makassar anchor Type 1

1942

UNWMKD

WMKD

JAPANESE NAVAL CONTROL AREA
(Old Dutch Indies)
JAPANESE OCCUPATION ISSUES

South Celebes
Makassar anchor Type 1

  I   I

          blue

131

NAVAL CONTROL AREA

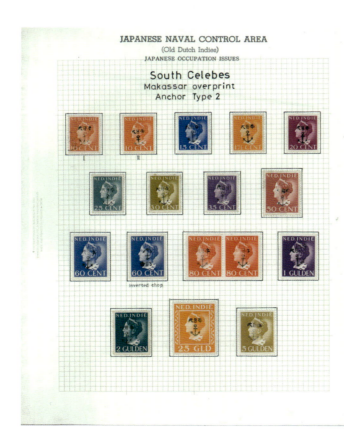
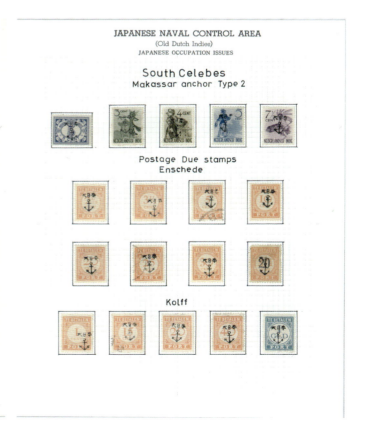

NAVAL CONTROL AREA

JAPANESE NAVAL CONTROL AREA
(Old Dutch Indies)
JAPANESE OCCUPATION ISSUES

South Celebes Anchor used in Borneo

Date 27 Oct.1944
Balikpapan

Date 22 Nov.1944 Bandjermasin. Anchor type 1

JAPANESE NAVAL CONTROL AREA
(Old Dutch Indies)
JAPANESE OCCUPATION ISSUES

South Celebes
Makassar anchor Type 2

UNWMKD

red & black double ovpt

WMKD

17
18

JAPANESE NAVAL CONTROL AREA
(Old Dutch Indies)
JAPANESE OCCUPATION ISSUES

South Celebes
Makassar overprint

Type 2 on covers

From Makassar.
To Malili.

Date May.1943 From Makassar. to Palopo.

JAPANESE NAVAL CONTROL AREA
(Old Dutch Indies)
JAPANESE OCCUPATION ISSUES

South Celebes
Makassar anchor Type 2

Date 27 Jan.1945.
From Makassar.
to Magelang.

From Madjene to Pare-Pare

133

NAVAL CONTROL AREA

JAPANESE NAVAL CONTROL AREA
(Old Dutch Indies)
JAPANESE OCCUPATION ISSUES

South Celebes
Makassar anchor Type 3

UNWMKD

WMKD

JAPANESE NAVAL CONTROL AREA
(Old Dutch Indies)
JAPANESE OCCUPATION ISSUES

South Celebes
Makassar anchor Type 3

I   II

JAPANESE NAVAL CONTROL AREA
(Old Dutch Indies)
JAPANESE OCCUPATION ISSUES

South Celebes
Makassar anchor Type 4

I   II

Netherlands

JAPANESE NAVAL CONTROL AREA
(Old Dutch Indies)
JAPANESE OCCUPATION ISSUES

South Celebes
Makassar anchor type 4

Date 11 May.1945 From Bonthain. To Malili.

Date 11 Jan 1945 From Makassar to Malili.

JAPANESE OCCUPATION "BALL FURONG" COLLECTION

NAVAL CONTROL AREA

JAPANESE NAVAL CONTROL AREA
(Old Dutch Indies)
JAPANESE OCCUPATION ISSUES

South Celebes
Makassar anchor Type 3
Postage Due stamps
Enschede

Kolff

JAPANESE NAVAL CONTROL AREA
(Old Dutch Indies)
JAPANESE OCCUPATION ISSUES

South Celebes
Makassar anchor Type 4
UNWMKD

WMKD

JAPANESE NAVAL CONTROL AREA
(Old Dutch Indies)
JAPANESE OCCUPATION ISSUES

South Celebes
Makassar anchor Type 5
UNWMKD

WMKD

JAPANESE NAVAL CONTROL AREA
(Old Dutch Indies)
JAPANESE OCCUPATION ISSUES

South Celebes
Makassar anchor Type 5

Type 6

Type 7

135

NAVAL CONTROL AREA

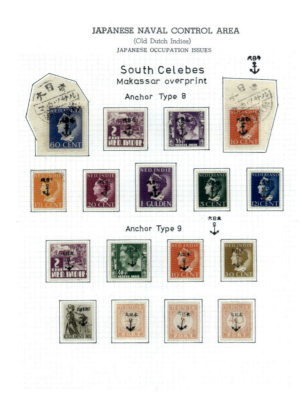

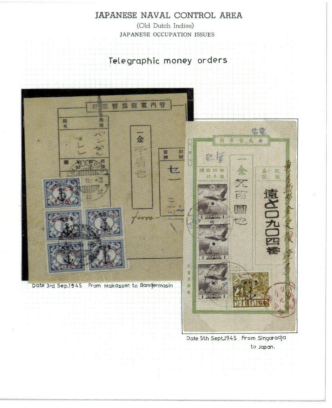

JAPANESE OCCUPATION "BALL FURONG" COLLECTION

NAVAL CONTROL AREA

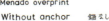

NAVAL CONTROL AREA

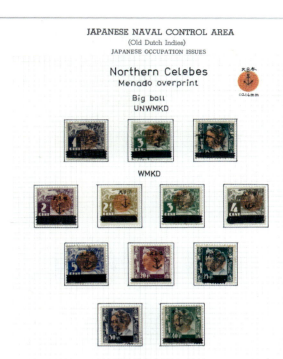

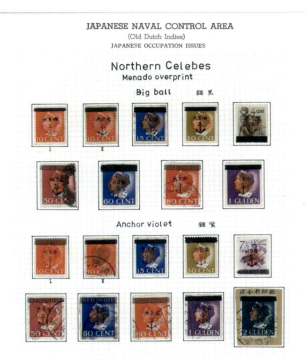

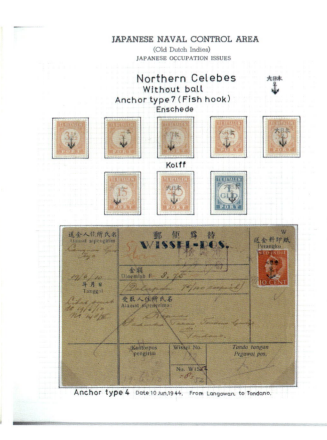

JAPANESE OCCUPATION "BALL FURONG" COLLECTION

NAVAL CONTROL AREA

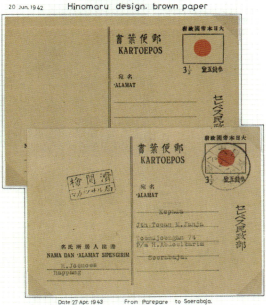

139

NAVAL CONTROL AREA

JAPANESE OCCUPATION "BALL FURONG" COLLECTION

NAVAL CONTROL AREA

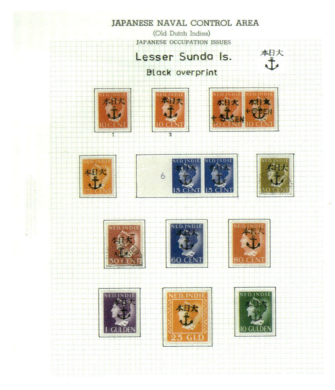

141

NAVAL CONTROL AREA

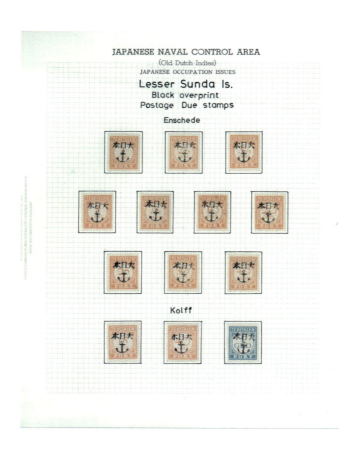

JAPANESE OCCUPATION "BALL FURONG" COLLECTION

NAVAL CONTROL AREA

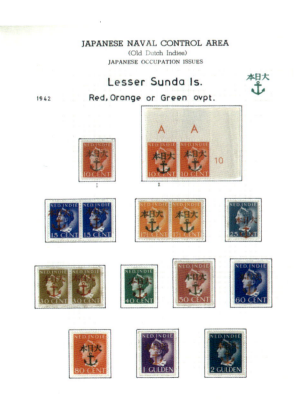

NAVAL CONTROL AREA

From Ambon, to Soerabaja.　アンボン から スラバヤ あて.

Ambon anchor　Typy 1

Black anchor

Red lilac anchor

Ambon anchor　Typy 2

144

JAPANESE OCCUPATION "BALL FURONG" COLLECTION

NAVAL CONTROL AREA

NAVAL CONTROL AREA

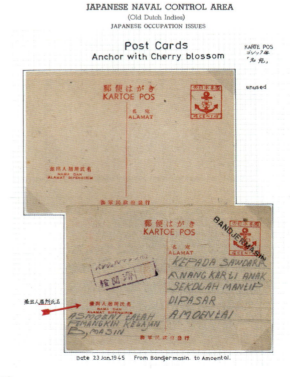

JAPANESE OCCUPATION "BALL FURONG" COLLECTION

NAVAL CONTROL AREA

147

NAVAL CONTROL AREA

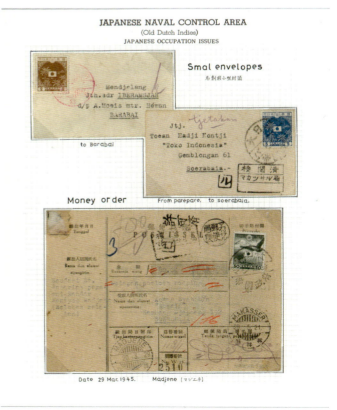

JAPANESE OCCUPATION "BALL FURONG" COLLECTION

NAVAL CONTROL AREA

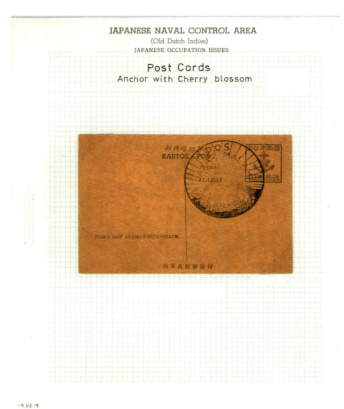

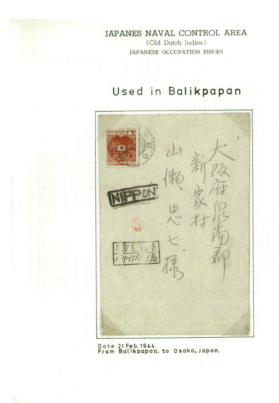

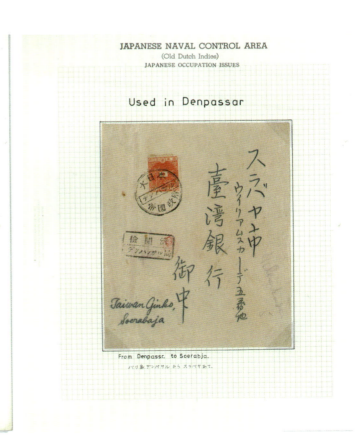

149

南方占領地のフィラテリー　玉芙蓉コレクション

NAVAL CONTROL AREA

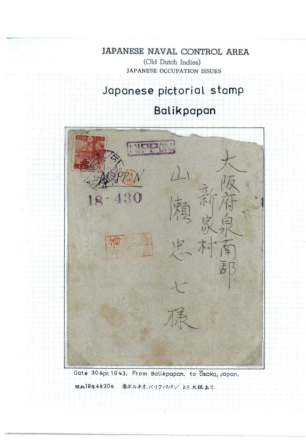

JAPANESE OCCUPATION "BALL FURONG" COLLECTION

NAVAL CONTROL AREA

JAPANESE NAVAL CONTROL AREA
(Old Dutch Indies)
JAPANESE OCCUPATION ISSUES

Japanese Pictorial Stamps
Used in Makassar

マカッサルで使用された日本切手

From Makassar to Tokyo, Japan. マカッサルより東京あて

JAPANESE NAVAL CONTROL AREA
(Old Dutch Indies)
JAPANESE OCCUPATION ISSUES

Japanese Pictorial stamp
Kolaka (south-celebes)

南セレベス コラカ から マラン (ジャワ島) あて

JAPANES NAVAL CONTROL AREA
(Old Dutch Indies)
JAPANESE OCCUPATION ISSUES

Post Cards of Japan
Surcharged in Black

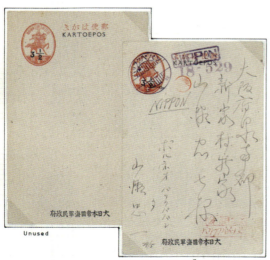

Unused

From Balikpapan to Ōsaka, Japan.
Date 29 May 1943
日付書込式特殊消印 ベリックパパン局

JAPANESE OCCUPATION ISSUES

"民政部" Overprinted

151

NAVAL CONTROL AREA

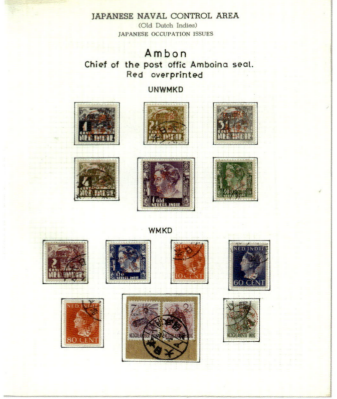

NAVAL CONTROL AREA

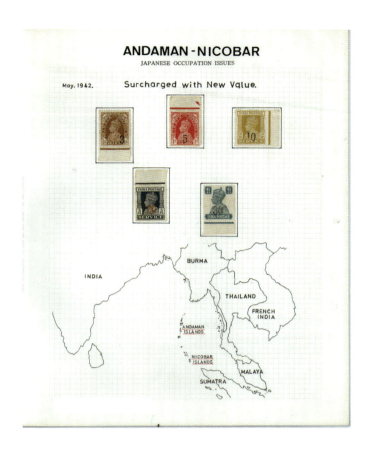